T0218321

SpringerBriefs in Statistics

JSS Research Series in Statistics

The current research of statistics in Japan has expanded in several directions in line with recent trends in academic activities in the area of statistics and statistical sciences over the globe. The core of these research activities in statistics in Japan has been the Japan Statistical Society (JSS). This society, the oldest and largest academic organization for statistics in Japan, was founded in 1931 by a handful of pioneer statisticians and economists and now has a history of about 80 years. Many distinguished scholars have been members, including the influential statistician Hirotugu Akaike, who was a past president of JSS, and the notable mathematician Kiyosi Itô, who was an earlier member of the Institute of Statistical Mathematics (ISM), which has been a closely related organization since the establishment of ISM. The society has two academic journals: the Journal of the Japan Statistical Society (English Series) and the Journal of the Japan Statistical Society (Japanese Series). The membership of JSS consists of researchers, teachers, and professional statisticians in many different fields including mathematics, statistics, engineering, medical sciences, government statistics, economics, business, psychology, education, and many other natural, biological, and social sciences. The JSS Series of Statistics aims to publish recent results of current research activities in the areas of statistics and statistical sciences in Japan that otherwise would not be available in English; they are complementary to the two JSS academic journals, both English and Japanese. Because the scope of a research paper in academic journals inevitably has become narrowly focused and condensed in recent years, this series is intended to fill the gap between academic research activities and the form of a single academic paper. The series will be of great interest to a wide audience of researchers, teachers, professional statisticians, and graduate students in many countries who are interested in statistics and statistical sciences, in statistical theory, and in various areas of statistical applications.

Yuzo Maruyama · Tatsuya Kubokawa ·
William E. Strawderman

Stein Estimation

 Springer

Yuzo Maruyama
Graduate School of Business
Administration
Kobe University
Kobe, Hyogo, Japan

Tatsuya Kubokawa
Faculty of Economics
University of Tokyo
Tokyo, Japan

William E. Strawderman
Department of Statistics
Rutgers University
Piscataway, NJ, USA

ISSN 2191-544X ISSN 2191-5458 (electronic)
SpringerBriefs in Statistics
ISSN 2364-0057 ISSN 2364-0065 (electronic)
JSS Research Series in Statistics
ISBN 978-981-99-6076-7 ISBN 978-981-99-6077-4 (eBook)
https://doi.org/10.1007/978-981-99-6077-4

This Springer imprint is published by the registered company Springer Nature Singapore Pte Ltd.
The registered company address is: 152 Beach Road, #21-01/04 Gateway East, Singapore 189721, Singapore

Paper in this product is recyclable.

Preface

This book provides a self-contained introduction to Stein's estimation, mainly focusing on minimaxity and admissibility. For estimation of a P-variate normal mean, if $X \sim \mathcal{N}_p(\mu, I)$, the estimator X is the MLE, UMVUE and is minimax with the constant risk p under the standard quadratic loss function. Stein (1956) showed that estimators of the form $(1 - a/(b + \|X\|^2))X$ dominate X, for a sufficiently small and b sufficiently large when $p \geq 3$. James and Stein (1961) explicitly constructed the dominating estimator, $(1 - (p - 2)/\|X\|^2)X$. Paradoxically the James-Stein estimator is itself inadmissible and can be dominated by another inadmissible estimate, its positive part. In such a situation, we are particularly interested in finding admissible estimators dominating a given inadmissible estimator.

Over 50 years ago, the third author of this book, Strawderman (1971) (WS) first constructed a class of proper Bayes admissible minimax estimators improving on X for $p \geq 5$. In that same year, Brown (1971) proposed sufficient conditions for a generalized Bayes estimator to be admissible. Brown's (1971) results led to a greatly enlarged class of generalized Bayes admissible minimax estimators.

Twenty years after Strawderman (1971), the second author of this book, Kubokawa (1991) (TK) first found an admissible estimator dominating the James-Stein estimator among the enlarged class of admissible minimax estimators. Further Kubokawa (1994) proposed sufficient condition for a shrinkage estimator to be superior to the James-Stein estimator. In that same year, 1994, the first author, Maruyama (YM), entered graduate school at the University of Tokyo, where the second author, (TK) was affiliated. YM got very interested in statistical decision theory, chose the second author, TK, as his supervisor, and wrote a Ph.D. thesis in this area. Theorems 2.7 and 3.15 are from that thesis.

After YM graduated, he contacted WS, arranged a visit to WS's University (Rutgers), and began a fruitful, continuing collaboration, resulting in several co-authored papers. Particularly, over the past five years, they have worked on admissible estimation of a multivariate normal mean when the scale is unknown. Recently, Maruyama and Strawderman (2021) constructed a class of generalized Bayes admissible minimax estimators for this longstanding open problem. Coincidentally, this occurred exactly 50 years after Strawderman (1971).

TK and WS also have had a longstanding collaboration resulting in several publications on minimaxity and admissibility in the general area covered by this book.

However, this thin book represents the first project on which the three of us have collaborated. It is our attempt to present a (nearly) self-contained introduction to Stein estimation. We focus on minimaxity and admissibility results in the estimation of the mean vector of a normal distribution in the known and unknown scale case when the covariance matrix is a multiple of the identity matrix and the loss is scaled squared error. For the most part, the estimators we study are spherically symmetric shrinkage estimators, often corresponding to generalized Bayes estimators relative to spherically symmetric (in μ) priors.

Due to space limitations, we do not cover the rich literature of estimation in restricted parameter spaces; estimation of loss (other than the SURE); loss functions or priors that are not spherically symmetric; or predictive density estimation. In particular, restricting attention to the spherically symmetric case allows a (relatively) direct and compact treatment of admissibility, and avoids the necessity to develop the deeper and more general, but somewhat more complex, development of Brown (1971).

We thank Yasuyuki Hamura for his helpful and constructive comments on an earlier version of this book. We are grateful to our families who have supported us during the preparation of this book.

Kobe, Japan   Yuzo Maruyama
Tokyo, Japan   Tatsuya Kubokawa
Piscataway, NJ, USA   William E. Strawderman
June 2023

Contents

1	**The Stein Phenomenon**	1
	1.1 Problem Setting	1
	1.2 Some Concepts of Statistical Decision Theory	2
	1.3 The Organization of This Book	6
	1.4 Stein Identity	9
	1.5 Preliminary Results: The Known Scale Case	12
	1.6 Preliminary Results: The Unknown Scale Case	17
	References	21
2	**Estimation of a Normal Mean Vector Under Known Scale**	23
	2.1 Introduction	23
	2.2 Review of Admissibility/Inadmissibility Results	25
	2.3 Inadmissibility	26
	2.4 Admissibility	29
	2.4.1 The Bayes Risk Difference	29
	2.4.2 A General Admissibility Result for Mixture Priors	30
	2.4.3 On the Boundary Between Admissibility and Inadmissibility	32
	2.5 Minimaxity	33
	2.5.1 $\xi(g)$ Bounded Near the Origin	34
	2.5.2 $\xi(g)$ Unbounded Near the Origin	37
	2.6 Improvement on the James–Stein Estimator	39
	References	42
3	**Estimation of a Normal Mean Vector Under Unknown Scale**	45
	3.1 Equivariance	45
	3.2 Proper Bayes Equivariant Estimators	47
	3.3 Admissible Bayes Equivariant Estimators Through the Blyth Method	52
	3.3.1 A General Admissibility Equivariance Result for Mixture Priors	53

3.3.2   On the Boundary Between Equivariant Admissibility
        and Inadmissibility 56
3.4   Admissibility Among All Estimators 57
      3.4.1   The Main Result 57
      3.4.2   A Proof of Theorem 3.6 59
3.5   Simple Bayes Estimators 62
3.6   Inadmissibility 63
      3.6.1   A General Sufficient Condition for Inadmissibility 63
      3.6.2   Inadmissible Generalized Bayes Estimators 66
3.7   Minimaxity 67
      3.7.1   A Sufficient Condition for Minimaxity 67
      3.7.2   Minimaxity of Some Generalized Bayes Estimators 70
3.8   Improvement on the James–Stein Estimator 71
References 76

Appendix A: Miscellaneous Lemmas and Technical Proofs 77

Chapter 1
The Stein Phenomenon

1.1 Problem Setting

In this book, we consider estimation of the mean vector of a multivariate normal distribution. Specifically, we consider i.i.d. p-variate normal random variables $Z_1, Z_2, \ldots, Z_m \sim \mathcal{N}_p(\mu, \sigma^2 I)$, where the mean vector μ is to be estimated. The sample mean vector is given by

$$X = \bar{Z} = \frac{1}{m} \sum_{i=1}^{m} Z_i \sim \mathcal{N}_p(\mu, I/\eta), \tag{1.1}$$

where $\eta = m/\sigma^2$. When the scale is known, we can set $\eta = 1$ without the loss of generality, that is,

$$X \sim \mathcal{N}_p(\mu, I),$$

where X is a complete sufficient statistic for μ. When σ^2, or equivalently η, is unknown, $\{X, S\}$ is a complete sufficient statistic for $\{\mu, \eta\}$, where

$$S = \frac{1}{m} \sum_{i=1}^{m} \|Z_i - \bar{Z}\|^2 \sim \frac{\chi_n^2}{\eta}, \quad \text{for } \eta = \frac{m}{\sigma^2} \text{ and } n = p(m-1). \tag{1.2}$$

Note that X and S are mutually independent and that η is a nuisance parameter. We will consider the known scale case and the unknown scale case, in Chaps. 2 and 3, respectively.

Much of this book is devoted to investigation of decision-theoretic properties of estimators in the two cases. While there are many similarities in the results, there are also some important differences. In order to help clarify the differences, for the unknown scale case, the mean vector μ is replaced by θ, and we use following notation:

© The Author(s), under exclusive license to Springer Nature Singapore Pte Ltd. 2023
Y. Maruyama et al., *Stein Estimation*, JSS Research Series in Statistics,
https://doi.org/10.1007/978-981-99-6077-4_1

$$X \sim \mathcal{N}_p(\theta, I/\eta) \text{ and } \eta S \sim \chi_n^2$$

in (1.1) and (1.2). In each case the loss function for estimation is assumed to be the scaled quadratic loss function;

$$L(\delta; \mu) = \|\delta(x) - \mu\|^2, \quad L(\delta; \theta, \eta) = \eta \|\delta(x, s) - \theta\|^2, \quad (1.3)$$

respectively.

1.2   Some Concepts of Statistical Decision Theory

In this section, we review some concepts of statistical decision theory, primarily for the known scale case. Denote the probability density function of $\mathcal{N}_p(\mu, I/\eta)$ by

$$\phi(x - \mu; \eta) = \frac{\eta^{p/2}}{(2\pi)^{p/2}} \exp\left(-\frac{\eta \|x - \mu\|^2}{2}\right).$$

We simply write $\phi(x - \mu)$ when $\eta = 1$.

The risk function of an estimator $\delta(x)$ is given by

$$R(\delta; \mu) = E[\|\delta(X) - \mu\|^2] = \int_{\mathbb{R}^p} \|\delta(x) - \mu\|^2 \phi(x - \mu) dx.$$

We say that an estimator δ dominates δ_0 if

$$
\begin{aligned}
R(\delta; \mu) &\le R(\delta_0; \mu) \text{ for all values of } \mu, \\
R(\delta; \mu) &< R(\delta_0; \mu) \text{ for at least one value of } \mu.
\end{aligned}
\qquad (1.4)
$$

Any estimator (such as δ_0 is in (1.4)) which is dominated by another estimator is said to be **inadmissible**. An estimator δ is **admissible** if it is not dominated by any other estimator. Admissibility is a relatively weak optimality property, while inadmissibility is often a compelling reason not to use a particular estimator.

An estimator δ_M of μ, which minimizes the maximum risk,

$$\sup_{\mu \in \mathbb{R}^p} R(\delta_M; \mu) = \inf_{\delta} \sup_{\mu \in \mathbb{R}^p} R(\delta; \mu)$$

is called a **minimax** estimator.

Bayes and **generalized Bayes** estimators also play an important roll in our development. For a proper prior density $\pi(\mu)$ which satisfies $\int_{\mathbb{R}^p} \pi(\mu) d\mu = 1$, (but for most of the development, $\int_{\mathbb{R}^p} \pi(\mu) d\mu < \infty$ suffices), the Bayes risk is defined by

$$r(\delta; \pi) = \int_{\mathbb{R}^p} R(\delta; \mu)\pi(\mu)d\mu.$$

Let the marginal density of x and the posterior density of μ given x be

$$m(x) = \int_{\mathbb{R}^p} \phi(x - \mu)\pi(\mu)d\mu \text{ and } \pi(\mu \mid x) = \frac{\phi(x - \mu)\pi(\mu)}{m(x)}, \quad (1.5)$$

respectively. Then the Bayes risk of $\delta(x)$, $r(\delta; \pi)$, is defined as

$$r(\delta; \pi) = \int_{\mathbb{R}^p} \left\{ \int_{\mathbb{R}^p} \|\delta(x) - \mu\|^2 \phi(x - \mu)dx \right\} \pi(\mu)d\mu$$

$$= \int_{\mathbb{R}^p} \left\{ \int_{\mathbb{R}^p} \|\delta(x) - \mu\|^2 \pi(\mu \mid x)d\mu \right\} m(x)dx.$$

The minimizer of $r(\delta; \pi)$ with respect to δ is called the **Bayes estimator** under π, and is given uniquely (a.e.) by

$$\delta_\pi(x) = \arg\min_d \int_{\mathbb{R}^p} \|d - \mu\|^2 \pi(\mu \mid x)d\mu \quad (1.6)$$

$$= \int_{\mathbb{R}^p} \mu\pi(\mu \mid x)d\mu = \frac{\int_{\mathbb{R}^p} \mu\phi(x - \mu)\pi(\mu)d\mu}{\int_{\mathbb{R}^p} \mu\phi(x - \mu)\pi(\mu)d\mu}.$$

Further, from (1.6), the equality

$$r(\delta; \pi) - r(\delta_\pi; \pi) = \int_{\mathbb{R}^p} \left\{ \int_{\mathbb{R}^p} \|\delta(x) - \delta_\pi(x)\|^2 \pi(\mu \mid x)d\mu \right\} m(x)dx \quad (1.7)$$

follows. By (1.7), we have the following result.

Theorem 1.1 *The proper Bayes estimator δ_π given by (1.6) is admissible.*

Proof Suppose the Bayes estimator δ_π is inadmissible, that is, there exists an estimator δ such that $\delta \neq \delta_\pi$ on a set of positive measure and

$$R(\delta; \mu) \leq R(\delta_\pi; \mu) \text{ for all values of } \mu.$$

Then

$$\int_{\mathbb{R}^p} \left\{ R(\delta_\pi; \mu) - R(\delta; \mu) \right\} \pi(\mu)d\mu = r(\delta_\pi; \pi) - r(\delta; \pi) \geq 0,$$

which contradicts (1.7), by uniqueness of δ_π.   □

Even if the prior is improper, $\int_{\mathbb{R}^p} \pi(\mu)d\mu = \infty$, the posterior density given by (1.5) is typically well-defined and hence the minimizer of (1.6) is also well-defined. The minimizer of (1.6) under an improper prior is called a **generalized Bayes estimator**. Unlike Theorem 1.1, a generalized Bayes estimator is not necessarily admissible. Parameter dimension often play a critical role in the determination of admissibility. A striking example of the effect of parameter dimension on admissibility is that of estimating a normal mean vector, which we tackle in this book. The natural estimator X, which is the maximum likelihood estimator and the uniformly minimum variance unbiased estimator, is also the generalized Bayes estimator with respect to the (improper) uniform prior on $\mathbb{R}^p$, since

$$\frac{\int_{\mathbb{R}^p} \mu\phi(x-\mu)d\mu}{\int_{\mathbb{R}^p} \phi(x-\mu)d\mu} = x + \frac{\int_{\mathbb{R}^p}(\mu-x)\phi(x-\mu)d\mu}{\int_{\mathbb{R}^p}\phi(x-\mu)d\mu} = x.$$

As noted in Theorems 1.4 and 1.7, X is admissible for $p = 1, 2$. However, for $p \geq 3$, it is inadmissible (Stein 1956, Theorems 1.5 and 1.8). This effect is called the **Stein phenomenon** since it was not at all expected at the time of its discovery.

Blyth (1951) method, given in Theorem 1.2 below, is applied to establish admissibility of a class of generalized Bayes estimators. Suppose $\pi(\mu)$ is improper,

$$\pi(\mu) > 0 \text{ for all } \mu, \tag{1.8}$$

and δ_π is the generalized Bayes estimator under $\pi(\mu)$. Suppose $\pi_i(\mu), i = 1, 2, \ldots$, is an increasing (in i) sequence of proper priors such that

$$\lim_{i \to \infty} \pi_i(\mu) = \pi(\mu) \text{ and } \pi_1(\mu) > 0 \text{ for all } \mu. \tag{1.9}$$

Note that the integral $\int_{\mathbb{R}^p} \pi_i(\mu)d\mu$ is not necessarily equal to 1, but must be finite for each i. Let δ_i be the proper Bayes estimator under π_i. Then the Bayes risk difference between δ_π and δ_i, with respect to $\pi_i(\mu)$, is

$$\Delta_i = \int_{\mathbb{R}^p} \{R(\delta_\pi; \mu) - R(\delta_i; \mu)\} \pi_i(\mu)d\mu. \tag{1.10}$$

The following form of Blyth's sufficient condition shows that $\lim_{i\to\infty} \Delta_i = 0$ implies admissibility.

Theorem 1.2 *The generalized Bayes estimator δ_π is admissible if* $\lim_{i\to\infty} \Delta_i = 0$.

Proof Suppose that δ_π is inadmissible and hence that there exists a δ' which satisfies

$$\begin{aligned} R(\delta'; \mu) &\leq R(\delta_\pi; \mu) \text{ for all values of } \mu, \\ R(\delta'; \mu_0) &< R(\delta_\pi; \mu_0) \text{ for some } \mu_0. \end{aligned} \tag{1.11}$$

By (1.11), we have

$$\int_{\mathbb{R}^p} \|\delta_\pi(x) - \delta'(x)\|^2 \phi(x - \mu_0) dx > 0.$$

Further we have

$$\int_{\mathbb{R}^p} \|\delta_\pi(x) - \delta'(x)\|^2 \phi(x - \mu) dx = \int_{\mathbb{R}^p} \|\delta_\pi(x) - \delta'(x)\|^2 \frac{\phi(x - \mu)}{\phi(x - \mu_0)} \phi(x - \mu_0) dx.$$

Since the ratio $\phi(x - \mu)/\phi(x - \mu_0)$ is continuous in x and positive, it follows that

$$\int_{\mathbb{R}^p} \|\delta_\pi(x) - \delta'(x)\|^2 \phi(x - \mu) dx > 0$$

for all μ. Set $\delta'' = (\delta_\pi + \delta')/2$. Then we have

$$\|\delta'' - \mu\|^2 = \frac{\|\delta_\pi - \mu\|^2 + \|\delta' - \mu\|^2}{2} - \frac{\|\delta_\pi - \delta'\|^2}{4},$$

and

$$R(\delta''; \mu) = E\left[\|\delta'' - \mu\|^2\right] < (1/2) E\left[\|\delta' - \mu\|^2\right] + (1/2) E\left[\|\delta_\pi - \mu\|^2\right]$$
$$= \frac{1}{2}\left\{R(\delta'; \mu) + R(\delta_\pi; \mu)\right\} \leq R(\delta_\pi; \mu),$$

for all μ. Then we have

$$\Delta_i = \int_{\mathbb{R}^p} \left\{R(\delta_\pi; \mu) - R(\delta_i; \mu)\right\} \pi_i(\mu) d\mu \geq \int_{\mathbb{R}^p} \left\{R(\delta_\pi; \mu) - R(\delta''; \mu)\right\} \pi_i(\mu) d\mu$$
$$\geq \int_{\mathbb{R}^p} \left\{R(\delta_\pi; \mu) - R(\delta''; \mu)\right\} \pi_1(\mu) d\mu > 0,$$

which contradicts $\Delta_i \to 0$ as $i \to \infty$.   □

Remark 1.1 In our version of Blyth's lemma, we assume (1.8) and (1.9), that is, $\pi(\mu) > 0$ and $\pi_1(\mu) > 0$ for any fixed μ, which are satisfied by all priors we assume in this book. In fact, the theorem still is true with essentially the same proof with the weaker assumption that $\pi_1(C) > 0$ for some compact set C. See Theorem 1.1 of Fourdrinier et al. (2018) for details.

When we apply Blyth's method in proving admissibility of δ_π, in some cases including Theorem 1.4 with (1.33), we bound Δ_i from above as $\Delta_i < C/\log(i + 1)$ for some positive constant C, which immediately implies $\lim_{i \to \infty} \Delta_i = 0$. However, in most cases, the use of Blyth's method in proving admissibility of δ_π, consists of first noting that the integrand of Δ_i in (1.10) tends to 0 as i tends to infinity. The proof is completed by showing that the integrand is bounded by an integrable function. Then, by the dominated convergence theorem, $\lim_{i \to \infty} \Delta_i = 0$ is satisfied so that δ_π is admissible.

1.3 The Organization of This Book

In Sects. 1.5 and 1.6, we give some preliminary results for the known scale and unknown scale cases, respectively. In both cases, the natural estimator X with the constant risk, p, is shown to be minimax for any $p \in \mathbb{N}$ (Theorems 1.3 and 1.6). Further the estimator X is shown to be admissible for $p = 1, 2$ (Theorems 1.4 and 1.7). However, for $p \geq 3$, the James–Stein estimator

$$\hat{\mu}_{JS} = \left(1 - \frac{p-2}{\|X\|^2}\right)X, \quad \hat{\theta}_{JS} = \left(1 - \frac{p-2}{n+2}\frac{S}{\|X\|^2}\right)X, \tag{1.12}$$

dominates X for the known scale case and the unknown scale case, respectively, which implies that the estimator X is inadmissible in both cases (Theorems 1.5 and 1.8). In this section, we preview the principal admissibility/inadmissibility and minimaxity results, which will be given in Chaps. 2 and 3.

In Chap. 2, we consider admissibility, inadmissibility and minimaxity of (generalized) Bayes estimators for the known scale. In particular, we mainly focus on a class of the (generalized) Bayes estimators with respect to an extended (Strawderman 1971)-type prior

$$\pi(\|\mu\|^2) = \int_0^\infty \frac{g^{-p/2}}{(2\pi)^{p/2}} \exp\left(-\frac{\|\mu\|^2}{2g}\right)\pi(g; a, b, c)dg$$
$$\text{where } \pi(g; a, b, c) = \frac{1}{(g+1)^a}\left(\frac{g}{g+1}\right)^b \frac{1}{\{\log(g+1)+1\}^c}, \tag{1.13}$$
$$\text{for } a > -p/2+1, \ b > -1 \text{ and } c \in \mathbb{R}.$$

Strawderman (1971) original parameterization is $\lambda = 1/(g+1) \in (0, 1)$. In particular, the case $a = b = c = 0$ as well as $p \geq 3$ corresponds to the Stein (1974) prior

$$\pi_S(\|\mu\|^2) = \|\mu\|^{2-p} \tag{1.14}$$

since

$$\int_0^\infty \frac{g^{-p/2}}{(2\pi)^{p/2}} \exp\left(-\frac{\|\mu\|^2}{2g}\right)dg = \frac{\Gamma(p/2-1)2^{p/2-1}}{(2\pi)^{p/2}}\|\mu\|^{2-p}.$$

For the behavior of $\pi(g; a, b, c)$ around the origin, we have $\lim_{g \to 0} \pi(g; a, b, c)/g^b = 1$ and

$$\int_0^1 g^b dg = \frac{1}{b+1} < \infty \tag{1.15}$$

if $b > -1$, which we assume through this book. For the asymptotic behavior of $\pi(g; a, b, c)$, we have $\lim_{g \to \infty} \pi(g; a, b, c)(g+1)^a \{\log(g+1)\}^c = 1$ and

$$\int_1^\infty \frac{dg}{(g+1)^a \{\log(g+1)\}^c} < \infty \tag{1.16}$$

if

$$\text{either } \{a > 1, \ c \in \mathbb{R}\} \text{ or } \{a = 1, \ c > 1\}. \tag{1.17}$$

The integrability conditions (1.15) and (1.16) imply that $\pi(g; a, b, c)$ is proper if $b > -1$ and $\{a, c\}$ satisfy (1.17). As in (2.5) in Sect. 2.1, if

$$a > -p/2 + 1, \ b > -1 \text{ and } c \in \mathbb{R},$$

the posterior density (and hence) the corresponding generalized Bayes estimator is well-defined.

For the corresponding (generalized) Bayes estimator, we have the following results:

Admissibility (Corollary 2.2 and Theorem 2.5)

$$\text{provided } p \geq 1 \text{ and either } \{a > 0, \ c \in \mathbb{R}\} \text{ or } \{a = 0, \ c \geq -1\}, \tag{1.18}$$

Inadmissibility (Corollary 2.1)

$$\text{provided } p \geq 3 \text{ and either } \{-p/2 + 1 < a < 0, \ c \in \mathbb{R}\} \text{ or } \{a = 0, \ c < -1\},$$

Minimaxity (Corollary 2.3)

$$\text{provided } p \geq 3, \ -p/2 + 1 < a \leq p/2 - 1, \ b \geq 0, \ c \leq 0.$$

Minimaxity results, under $\{b \geq 0 \text{ and } c > 0\}$ and $\{-1 < b < 0 \text{ and } c = 0\}$, are also provided in Corollary 2.3 and Theorem 2.7, respectively. Further we find a generalized Bayes estimator improving on the James–Stein estimator $\hat{\mu}_{JS}$ given by (1.12).

Improving on the James–Stein estimator   (Theorem 2.8)

$$\text{provided } p \geq 3, \ a = b = c = 0. \tag{1.19}$$

A particularly interesting case when the scale is known seems deserving of attention. When $a = b = c = 0$, the prior corresponds to the Stein (1974) prior π_S given by (1.14). As in (1.3), (1.18) and (1.19), the corresponding generalized Bayes estimator is minimax, admissible, and, as first established by Kubokawa (1991), superior to the James–Stein estimator.

In Chap. 3, we treat the unknown scale case. We first consider admissibility within the class of (scale) equivariant estimators of the form $\{1 - \psi(\|x\|^2/s)\}x$. A joint prior of the form $\eta^\nu \eta^{p/2} q(\eta \|\theta\|^2)$ for $\nu \in \mathbb{R}$, gives rise to such an invariant generalized Bayes estimator. We show that the choice $\nu = -1$ is special in the sense that a joint prior of the form

$$\eta^{-1} \eta^{p/2} q(\eta \|\theta\|^2),$$

where $q(\|\mu\|^2)$ is proper on $\mu \in \mathbb{R}^p$, gives a generalized Bayes estimator which is admissible within the class of equivariant estimators. Further, we investigate the properties of the generalized Bayes estimators under $\eta^{-1} \eta^{p/2} \pi(\eta \|\theta\|^2)$ with $\pi(\cdot)$ by (1.13). Here are some of the main results:

Admissibility among all equivariant estimators   (Corollary 3.1 and Theorem 3.5)

$$\text{provided } p \geq 1 \text{ and either } \{a > 0, \ c \in \mathbb{R}\} \text{ or } \{a = 0, \ c > -1\}$$

Admissibility among all estimators   (Theorem 3.6)

$$\text{provided } p \geq 1, \ 0 < a < n/2 + 2, \ c = 0. \tag{1.20}$$

Inadmissibility   (Corollaries 3.3 and 3.4)

$$\text{provided } p \geq 3 \text{ and either } \{-p/2 + 1 < a < 0, \ c \in \mathbb{R}\} \text{ or } \{a = 0, \ c < -1\},$$

As in (1.18), the choice

$$\text{either } \{a > 0, \ c \in \mathbb{R}\} \text{ or } \{a = 0, \ c \geq -1\},$$

implies admissibility for the known scale case. We conjecture that the choice (1.18) implies admissibility for the unknown scale case as well.

We also find minimax generalized Bayes estimators:

Minimaxity   (Corollary 3.5)

$$\text{provided } p \geq 3, \ b \geq 0, \ c \leq 0, \ -p/2 + 1 < a \leq \frac{(p-2)(n+2)}{2(2p+n-2)}. \quad (1.21)$$

The minimaxity results, under $\{b \geq 0, c > 0\}$ and $\{-1 < b < 0, c = 0\}$ are also provided in Corollary 3.5 and Theorem 3.13, respectively. Further we find a generalized Bayes estimator improving on the James–Stein estimator $\hat{\theta}_{JS}$ given by (1.12).

Improving on the James–Stein estimator   (Theorem 3.14)

$$\text{provided } p \geq 3, \ a = b = c = 0. \quad (1.22)$$

Two interesting cases when the scale is unknown seem deserving of attention. When $a = b = c = 0$, the prior corresponds to the joint (Stein 1974) prior

$$\eta^{-1} \times \eta^{p/2} \pi_S(\eta \|\theta\|^2) = \eta^{-1} \times \eta^{p/2} \left\{\eta \|\theta\|^2\right\}^{1-p/2} = \|\theta\|^{2-p}, \quad (1.23)$$

where π_S is given by (1.14). As in (1.20), (1.21) and (1.22), the corresponding generalized Bayes estimator is admissible among all equivariant estimators, minimax, and superior to the the James–Stein estimator, respectively, where the last result was first established by Kubokawa (1991).

Another interesting case is a variant of the James–Stein estimator of the simple form

$$\left(1 - \frac{\alpha}{\|x\|^2/s + \alpha + 1}\right)x,$$

which is generalized Bayes under $\pi(g; a, b, c)$ with $c = 0$, $b = n/2 - a$, $\alpha = (p/2 - 1 + a)/(n/2 + 1 - a)$, as shown in Sect. 3.5. By (1.20) and (1.21), this estimator with $(p-2)/(n+2) < \alpha \leq 2(p-2)/(n+2)$ is minimax and admissible.

1.4 Stein Identity

The following identity is called the (one dimensional) (Stein 1974) identity (or Stein's lemma). The identity not only provides a much easier proof of the initial results on the Stein phenomenon, but also has been the most powerful technique for further developments in this area.

Lemma 1.1 (Stein 1974) *Let $X \sim \mathcal{N}(\theta, 1/\eta)$, and let $f(x)$ be a function such that*

$$f(b) - f(a) = \int_a^b f'(x)dx \tag{1.24}$$

for all $a, b \in \mathbb{R}$. Further suppose $\mathrm{E}[|f'(X)|] < \infty$. Then we have

$$\eta\,\mathrm{E}[(X - \theta)f(X)] = \mathrm{E}[f'(X)].$$

Proof We have

$$\eta\,\mathrm{E}[(X - \theta)f(X)] = \eta\,\mathrm{E}[(X - \theta)\{f(X) - f(\theta)\}] \tag{1.25}$$

$$= \eta\left(\int_\theta^\infty + \int_{-\infty}^\theta\right)\{f(x) - f(\theta)\}(x - \theta)\frac{\eta^{1/2}}{(2\pi)^{1/2}}\exp\left(-\frac{\eta(x - \theta)^2}{2}\right)dx$$

$$= \eta\int_\theta^\infty \int_\theta^x f'(y)(x - \theta)\frac{\eta^{1/2}}{(2\pi)^{1/2}}\exp\left(-\frac{\eta(x - \theta)^2}{2}\right)dydx$$

$$- \eta\int_{-\infty}^\theta \int_x^\theta f'(y)(x - \theta)\frac{\eta^{1/2}}{(2\pi)^{1/2}}\exp\left(-\frac{\eta(x - \theta)^2}{2}\right)dydx,$$

where the third equality follows from (1.24). Note

$$\{(x, y)\,|\,\theta < x < \infty,\ \theta < y < x\} = \{(x, y)\,|\,y < x < \infty, \theta < y < \infty\}, \tag{1.26}$$

for the first term of the right-hand side of (1.25) and

$$\{(x, y)\,|\,-\infty < x < \theta,\ x < y < \theta\}$$
$$= \{(x, y)\,|\,-\infty < x < y,\ -\infty < y < \theta\}, \tag{1.27}$$

for the second term of the right-hand side of (1.25). By (1.26), (1.27), and Fubini's theorem justifying the interchange of order of integration, we have

$$\eta \, \mathrm{E}[(X-\theta)f(X)]$$

$$= \int_\theta^\infty \int_y^\infty f'(y)\eta(x-\theta)\frac{\eta^{1/2}}{(2\pi)^{1/2}} \exp\left(-\frac{\eta(x-\theta)^2}{2}\right) dx\, dy$$

$$- \int_{-\infty}^\theta \int_{-\infty}^y f'(y)\eta(x-\theta)\frac{\eta^{1/2}}{(2\pi)^{1/2}} \exp\left(-\frac{\eta(x-\theta)^2}{2}\right) dx\, dy$$

$$= \int_\theta^\infty f'(y)\left\{\int_y^\infty \eta(x-\theta)\frac{\eta^{1/2}}{(2\pi)^{1/2}} \exp\left(-\frac{\eta(x-\theta)^2}{2}\right) dx\right\} dy$$

$$- \int_{-\infty}^\theta f'(y)\left\{\int_{-\infty}^y \eta(x-\theta)\frac{\eta^{1/2}}{(2\pi)^{1/2}} \exp\left(-\frac{\eta(x-\theta)^2}{2}\right) dx\right\} dy$$

$$= \left(\int_\theta^\infty + \int_{-\infty}^\theta\right) f'(y)\frac{\eta^{1/2}}{(2\pi)^{1/2}} \exp\left(-\frac{\eta(y-\theta)^2}{2}\right) dy$$

$$= \mathrm{E}[f'(X)],$$

which completes the proof. □

In higher dimensions, let $f = (f_1, \ldots, f_p)$ be a function from $\mathbb{R}^p$ into $\mathbb{R}^p$. Also, for any $x = (x_1, \ldots, x_p) \in \mathbb{R}^p$ and for fixed $j = 1, \ldots, p$, set $x_{-j} = (x_1, \ldots, x_{j-1}, x_j, \ldots, x_p)$, and, with a slight abuse of notation, $x = (x_j, x_{-j})$. Then, using the independence of X_i and X_{-j}, we have

$$\eta \, \mathrm{E}[(X_j - \mu_j)f_j(X)] = \mathrm{E}[\eta \, \mathrm{E}[(X_j - \mu_j)f_j(X_j, X_{-j})] \mid X_j]$$
$$= \mathrm{E}[\mathrm{E}[\{\partial/\partial x_j\} f_i(X_j, X_{-j})] \mid X_j] = \mathrm{E}[\{\partial/\partial x_j\} f_j(X)]. \quad (1.28)$$

A more careful treatment of (1.28) can be seen in Chap. 2 of Fourdrinier et al. (2018), where the development is extended to weakly differentiable functions which will be useful in the rest of this book.

Stein's lemma can also be used to establish a useful identity for Chi-square variables which are helpful in extending shrinkage estimation results from the known variance case to the unknown variance case. Let $Y_1, \ldots, Y_n$ be independent normal random variables with $Y_i \sim N(0, 1/\eta)$, and let $S = \sum_{i=1}^n Y_i^2$. Then $\eta S \sim \chi_n^2$. The following was proposed as the Chi-square identity by Efron and Morris (1976).

Lemma 1.2 (Chi-square identity) *Let $\eta S \sim \chi_n^2$. Then*

$$\eta \, \mathrm{E}[Sf(S)] = \mathrm{E}[nf(S) + 2Sf'(S)].$$

Proof Lemma 1.1 and (1.28) gives

$$\eta \mathrm{E}\left[Y_j^2 f\left(\sum_{j=1}^n Y_j^2\right)\right] = \eta \mathrm{E}\left[Y_j\left\{Y_j f\left(\sum_{j=1}^n Y_j^2\right)\right\}\right]$$

$$= \mathrm{E}\left[f\left(\sum_{j=1}^n Y_j^2\right) + 2Y_j^2 f'\left(\sum_{j=1}^n Y_j^2\right)\right].$$

Then we have

$$\eta \mathrm{E}\left[Sf(S)\right] = \eta \mathrm{E}\left[\sum_{j=1}^n Y_i^2 f\left(\sum_{j=1}^n Y_j^2\right)\right]$$

$$= \sum_{j=1}^n \mathrm{E}\left[f\left(\sum_{j=1}^n Y_j^2\right) + 2Y_j^2 f'\left(\sum_{j=1}^n Y_j^2\right)\right] = \mathrm{E}[nf(S) + 2Sf'(S)],$$

which completes the proof.   □

1.5 Preliminary Results: The Known Scale Case

In this section we consider admissibility and minimaxity of the estimator X for all dimensions, p, in the case of a known scale. We first address minimaxity.

Theorem 1.3 *The estimator X is minimax for all $p \in \mathbb{N}$.*

Proof Recall $\mathrm{R}(X; \mu) = \mathrm{E}[\|X - \mu\|^2] = p$. Consider the normal prior with mean 0 and covariance iI,

$$\pi_i(\mu) = \frac{1}{(2\pi i)^{p/2}} \exp\left(-\frac{\|\mu\|^2}{2i}\right). \tag{1.29}$$

Then, by Lemma A.1 and (1.6), the Bayes estimator under $\pi_i(\mu)$ is

$$\hat{\mu}_i(x) = \frac{\int_{\mathbb{R}^p} \mu \phi(x - \mu)\pi_i(\mu)\mathrm{d}\mu}{\int_{\mathbb{R}^p} \phi(x - \mu)\pi_i(\mu)\mathrm{d}\mu} = \frac{i}{i+1}x, \tag{1.30}$$

where the risk of $\hat{\mu}_i$ is given by

$$\mathrm{R}(\hat{\mu}_i; \mu) = \mathrm{E}\left[\|\hat{\mu}_i - \mu\|^2\right] = \left(\frac{i}{i+1}\right)^2 p + \left(\frac{1}{i+1}\right)^2 \|\mu\|^2. \tag{1.31}$$

For the prior (1.29), we have

$$\int_{\mathbb{R}^p} \|\mu\|^2 \pi_i(\mu)\mathrm{d}\mu = ip$$

and hence, with the risk (1.31), the Bayes risk of $\hat{\mu}_i$ under $\pi_i(\mu)$ is

$$r(\hat{\mu}_i, \pi_i) = \int_{\mathbb{R}^p} R(\hat{\mu}_i; \mu)\pi_i(\mu)d\mu = \left(\frac{i}{i+1}\right)p.$$

Therefore, for any estimator δ,

$$\sup_\mu R(\delta; \mu) \geq \int_{\mathbb{R}^p} R(\delta; \mu)\pi_i(\mu)d\mu = r(\delta, \pi_i) \geq r(\hat{\mu}_i, \pi_i) = \left(\frac{i}{i+1}\right)p,$$

for all $i \in \mathbb{N}$. Hence

$$\sup_\mu R(\delta; \mu) \geq p = R(X; \mu) = \sup_\mu R(X; \mu),$$

which completes the proof of minimaxity of X for any p.   □

The estimator X is the generalized Bayes estimator with respect to the improper uniform prior on $\mathbb{R}^p$. Since the estimator X is the maximum likelihood, and uniformly minimum variance unbiased estimator, and is also minimax, it is natural to expect that it is also admissible. It was, therefore, very surprising when Stein (1956) showed that X is inadmissible when $p \geq 3$. It is however, admissible for $p \leq 2$.

In order to establish the admissibility of the estimator X through the Blyth method, let

$$k_i(g) = 1 - \frac{\log(g+1)}{\log(g+1+i)}. \tag{1.32}$$

Clearly $k_i(g)$ is increasing in i for fixed g, and decreasing in g for fixed i. Further $\lim_{i\to\infty} k_i(g) = 1$ for fixed $g \geq 0$. Some properties of k_i will be given in Lemma A.6 in Sect. A.2. In particular, for $p = 1, 2$, Part 2 of Lemma A.6 guarantees the integrability

$$\int_{\mathbb{R}^p} k_i^2(\|\mu\|^2)d\mu = \frac{\pi^{p/2}}{\Gamma(p/2)} \int_0^\infty g^{p/2-1}k_i^2(g)dg$$
$$\leq \frac{\pi^{p/2}}{\Gamma(p/2)} \left(\int_0^1 g^{-1/2}dg + \int_0^\infty k_i^2(g)dg \right) \leq \frac{\pi^{p/2}}{\Gamma(p/2)}\{2 + (1+i)\} < \infty,$$

where the first equality follows from Part 1 of Lemma A.2. Using the sequence $k_i^2(\|\mu\|^2)$ for (1.32) and the Blyth method (Theorem 1.2), we have the following result.

Theorem 1.4 *The estimator X is admissible for $p = 1, 2$.*

Proof The Bayes estimator with respect to the prior density $k_i^2(\|\mu\|^2)$ is

$$\hat{\mu}_i(x) = \frac{\int_{\mathbb{R}^p} \mu\phi(x-\mu)k_i^2(\|\mu\|^2)d\mu}{\int_{\mathbb{R}^p} \phi(x-\mu)k_i^2(\|\mu\|^2)d\mu} = x + \frac{\int_{\mathbb{R}^p} (\mu-x)\phi(x-\mu)k_i^2(\|\mu\|^2)d\mu}{\int_{\mathbb{R}^p} \phi(x-\mu)k_i^2(\|\mu\|^2)d\mu}$$

$$= x + \frac{\int_{\mathbb{R}^p} \phi(x-\mu)\nabla\{k_i^2(\|\mu\|^2)\}d\mu}{\int_{\mathbb{R}^p} \phi(x-\mu)k_i^2(\|\mu\|^2)d\mu},$$

where the last equality follows from Lemma 1.1 and (1.28). By the identity

$$\|x-\mu\|^2 - \|\hat{\mu}_i - \mu\|^2 = -\|\hat{\mu}_i - x\|^2 + 2(\hat{\mu}_i - x)^{\mathsf{T}}(\mu - x),$$

the non-scaled Bayes risk difference is

$$\Delta_i = \int_{\mathbb{R}^p} \left(\mathrm{E}[\|X-\mu\|^2] - \mathrm{E}[\|\hat{\mu}_i(X)-\mu\|^2] \right) k_i^2(\|\mu\|^2)d\mu$$

$$= \int_{\mathbb{R}^p} \frac{\|\int_{\mathbb{R}^p} \phi(x-\mu)\nabla\{k_i^2(\|\mu\|^2)\}d\mu\|^2}{\int_{\mathbb{R}^p} \phi(x-\mu)k_i^2(\|\mu\|^2)d\mu}dx$$

$$= \int_{\mathbb{R}^p} \frac{\|4\int_{\mathbb{R}^p} \mu\phi(x-\mu)k_i(\|\mu\|^2)k_i'(\|\mu\|^2)d\mu\|^2}{\int_{\mathbb{R}^p} \phi(x-\mu)k_i^2(\|\mu\|^2)d\mu}dx.$$

By the Cauchy-Schwarz inequality (Part 2 of Lemma A.3) and Part 1 of Lemma A.2, we have

$$\Delta_i \leq 16 \int_{\mathbb{R}^p} \int_{\mathbb{R}^p} \phi(x-\mu)\|\mu\|^2\{k_i'(\|\mu\|^2)\}^2 d\mu dx$$

$$= 16\frac{\pi^{p/2}}{\Gamma(p/2)} \int_{\mathbb{R}^p} \phi(x)dx \int_0^\infty g^{p/2-1}g\{k_i'(g)\}^2 dg$$

$$= 16\frac{\pi^{p/2}}{\Gamma(p/2)} \int_0^\infty g^{p/2}\{k_i'(g)\}^2 dg.$$

Note, for $p = 1, 2$, $g^{p/2} \leq g + 1$ for all $g \geq 0$. Hence, by Part 4 of Lemma A.6,

$$\Delta_i \leq 16\frac{\pi^{p/2}}{\Gamma(p/2)} \int_0^\infty (g+1)\{k_i'(g)\}^2 dg \leq \frac{32\pi^{p/2}}{\Gamma(p/2)} \frac{1}{\log(1+i)}. \tag{1.33}$$

Thus we have $\Delta_i \to 0$ as $i \to \infty$, which completes the proof. $\square$

Remark 1.2 Blyth (1951) showed the admissibility of X for $p = 1$, by using the Gaussian sequence (1.29). Stein (1956) showed the admissibility of X for $p = 2$, not by the Blyth method, but by the information inequality method, mainly because the Blyth method, using the Gaussian sequence (1.29), does not work for $p = 2$. James and Stein (1961) showed the admissibility of X for $p = 2$ by the Blyth method with a proper sequence somewhat similar to k_i given by (1.32). Our sequence k_i, which is always strictly positive, is regarded as a more sophisticated version of the sequence of James and Stein (1961) as well as Brown and Hwang (1982), Maruyama and Takemura (2008) and Maruyama (2009).

We next address the inadmissibility of X for $p \geq 3$. The risk function of an estimator of the form $\delta_\psi(x) = \{1 - \psi(\|x\|^2)\}x$ is given by

$$R(\delta_\psi; \mu) = E\left[\|\delta_\psi(X) - \mu\|^2\right]$$

$$= E\left[\|X - \mu\|^2\right] + E\left[\|X\|^2 \psi^2(\|X\|^2)\right] - 2\sum_{i=1}^{p} E\left[(X_i - \mu_i)X_i\psi(\|X\|^2)\right].$$

$$(1.34)$$

Then the Stein identity (Lemma 1.1 and (1.28)) gives

$$E\left[(X_i - \mu_i)X_i\psi(\|X\|^2)\right] = E\left[\psi(\|X\|^2) + 2X_i^2\psi'(\|X\|^2)\right],$$

$$\text{and } \sum_{i=1}^{p} E\left[(X_i - \mu_i)X_i\psi(\|X\|^2)\right] = E\left[p\psi(\|X\|^2) + 2\|X\|^2\psi'(\|X\|^2)\right].$$

$$(1.35)$$

By (1.34) and (1.35), we have $R(\delta_\psi; \mu) = E\left[\hat{R}_\psi(\|X\|^2)\right]$, where $\hat{R}_\psi(\|x\|^2)$ is called the SURE (Stein Unbiased Risk Estimate) and is given by

$$\hat{R}_\psi(w) = p + w\psi^2(w) - 2p\psi(w) - 4w\psi'(w).$$

$$(1.36)$$

Now let $\psi_{JS}(w) = (p - 2)/w$ for $p \geq 3$, which corresponds to the James and Stein (1961) estimator,

$$\hat{\mu}_{JS} = \left(1 - \frac{p-2}{\|X\|^2}\right)X.$$

It follows that $\hat{R}_{JS}(w) = p - (p - 2)^2/w$, which implies that the risk of the estimator $\hat{\mu}_{JS}$ is

$$R(\hat{\mu}_{JS}; \mu) = E[\hat{R}_{JS}(\|X\|^2)] = p - E\left[\frac{(p-2)^2}{\|X\|^2}\right] \leq p = R(X; \mu). \quad (1.37)$$

Note that the above expectation exists and is finite for $p \geq 3$, but not for $p \leq 2$. Hence we have the following result.

Theorem 1.5 *The estimator X is inadmissible for $p \geq 3$.*

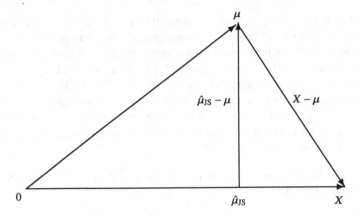

Fig. 1.1 Pythagorean triangle of the James–Stein estimator

Two popular interpretations of the James–Stein estimator $\hat{\mu}_{JS}$ are as follows.

Remark 1.3 A geometric interpretation of the James–Stein estimator was provided by Stein (1962) and explained in Brandwein and Strawderman (1990). In (1.37), we have $(p-2)^2 \mathrm{E}[1/\|X\|^2] = \mathrm{E}[\|X - \hat{\mu}_{JS}\|^2]$ and hence

$$\mathrm{E}[\|\hat{\mu}_{JS} - \mu\|^2] + \mathrm{E}[\|X - \hat{\mu}_{JS}\|^2] = \mathrm{E}[\|X - \mu\|^2]. \tag{1.38}$$

Let $f(x)$ and $g(x)$ be two p-variate vectors of functions of x. Suppose the inner product of $f(x)$ and $g(x)$ is defined by

$$\langle f, g \rangle_{\mathrm{E}} = \mathrm{E}[f(X)^{\mathsf{T}}g(X)] = \int_{\mathbb{R}^p} f(x)^{\mathsf{T}}g(x)\phi(x - \mu)\mathrm{d}x,$$

and the norm of f is given by $\|f\|_{\mathrm{E}} = \sqrt{\langle f, f \rangle_{\mathrm{E}}}$. Then, as in Fig. 1.1, (1.38) can be written as

$$\langle X - \mu \rangle_{\mathrm{E}}^2 = \langle \hat{\mu}_{JS} - \mu \rangle_{\mathrm{E}}^2 + \langle X - \hat{\mu}_{JS} \rangle_{\mathrm{E}}^2,$$

which is a Pythagorean triangle among X, μ and $\hat{\mu}_{JS}$.

Remark 1.4 A motivation of the James–Stein estimator as an empirical Bayes estimator is given by Efron and Morris (1972a, b). The (empirical) Bayesian framework, uses the prior $\mu \sim \mathcal{N}_p(0, gI)$, with the variance g of the prior distribution treated as an unknown parameter. By (1.5), the marginal distribution of X and the posterior distribution of μ given $X = x$ are

$$X \sim \mathcal{N}_p(0, (g+1)I), \text{ and } \mu \,|\, x \sim \mathcal{N}_p(\hat{\mu}_{\mathrm{B}}, \{g/(g+1)\}I), \tag{1.39}$$

respectively, where $\hat{\mu}_B$ is the posterior mean or the Bayes estimator under quadratic loss, given by

$$\hat{\mu}_B = E[\mu \mid X = x] = x - \frac{1}{g+1}x. \tag{1.40}$$

Since g is unknown, we estimate $1/(g+1)$ in (1.40) from the marginal distribution given by (1.39). Noting that $\|X\|^2/(1+g)$ is distributed as χ_p^2, we have

$$E\left[\frac{1+g}{\|X\|^2}\right] = \frac{1}{p-2} \text{ or } E\left[\frac{p-2}{\|X\|^2}\right] = \frac{1}{1+g},$$

which implies that $(p-2)/\|X\|^2$ is an unbiased estimator of $1/(1+g)$. Substituting the unbiased estimator into the Bayes estimator $\hat{\mu}_B$ gives the empirical Bayes estimator which is identical to the James–Stein estimator.

As shown in Baranchik (1964), the James–Stein estimator is inadmissible since it is dominated by the James–Stein positive-part estimator

$$\hat{\mu}_{JS}^+ = \max\left(0, 1 - \frac{p-2}{\|X\|^2}\right)X.$$

In Sect. 2.6, we consider improving on the James–Stein estimator.

As mentioned above, this book focuses on the estimation of normal means, primarily in the case of a covariance matrix equal to a multiple of the identity. Many additional related problems on shrinkage/Stein estimation have been studied in the literature. Fourdrinier et al. (2018) covers some of these additional problems.

1.6 Preliminary Results: The Unknown Scale Case

In this section we consider admissibility and minimaxity of the estimator X for all dimensions, p, in the case of an unknown scale.

We first address minimaxity. In the unknown scale case, as in (1.3), the loss is scaled as

$$L(\delta; \theta, \eta) = \eta \|\delta(x, s) - \theta\|^2.$$

Let the density of $S \sim \chi_n^2/\eta$ be denoted by

$$f_n(s; \eta) = \frac{\eta^{n/2} s^{n/2-1}}{\Gamma(n/2) 2^{n/2}} \exp\left(-\frac{\eta s}{2}\right).$$

Then, for the joint prior $\pi(\theta, \eta)$, the Bayes estimator minimizes the Bayes risk given by

$$r(\delta; \pi) = \iint \left\{ \iint \eta \|\delta(x, s) - \theta\|^2 \phi(x - \theta; \eta) f_n(s; \eta) dx ds \right\} \pi(\theta, \eta) d\theta d\eta$$

$$= \iint \left\{ \iint \|\delta(x, s) - \theta\|^2 \tilde{\pi}(\theta, \eta \,|\, x, s) d\theta d\eta \right\} \tilde{m}(x, s) dx ds,$$

where

$$\tilde{m}(x, s) = \iint \eta \phi(x - \theta; \eta) f_n(s; \eta) \pi(\theta, \eta) d\theta d\eta,$$

$$\text{and } \tilde{\pi}(\theta, \eta \,|\, x, s) = \frac{\eta \phi(x - \theta; \eta) f_n(s; \eta) \pi(\theta, \eta)}{\tilde{m}(x, s)}.$$

Then, as in (1.6), the (proper) Bayes estimator, the minimizer of $r(\delta; \pi)$, is given by

$$\delta_\pi(x, s) = \iint \theta \tilde{\pi}(\theta, \eta \,|\, x, s) d\theta d\eta = \frac{\iint \eta \theta \phi(x - \theta; \eta) \pi(\theta, \eta) d\theta d\eta}{\iint \eta \phi(x - \theta; \eta) \pi(\theta, \eta) d\theta d\eta}, \quad (1.41)$$

which is not the posterior mean of θ. Generalized Bayes estimators have the same form provided the above integrals exist and are finite.

Theorem 1.6 *The estimator X is minimax for all $p \in \mathbb{N}$.*

Proof Recall $R(X; \theta, \eta) = E[\eta \|X - \theta\|^2] = p$. Assume the prior on (θ, η) with

$$\pi_i(\theta \,|\, \eta) \times \pi(\eta), \quad (1.42)$$

where $\pi(\eta)$ is any proper prior and

$$\pi_i(\theta \,|\, \eta) = \frac{\eta^{p/2}}{(2\pi i)^{p/2}} \exp\left(-\frac{\eta \|\theta\|^2}{2i}\right).$$

As in (1.30), Lemma A.1 and (1.41) give the Bayes estimator under the prior (1.42), as

$$\hat{\theta}_i = \frac{\iint \eta \theta \phi(x - \theta; \eta) f_n(s; \eta) \pi_i(\theta \,|\, \eta) \pi(\eta) d\theta d\eta}{\iint \eta \phi(x - \theta; \eta) f_n(s; \eta) \pi_i(\theta \,|\, \eta) \pi(\eta) d\theta d\eta} = \frac{i}{i+1} x.$$

Further, as in (1.31), the risk of $\hat{\theta}_i$ is

$$R(\hat{\theta}_i; \theta, \eta) = E\left[\eta \|\hat{\theta}_i - \theta\|^2\right] = \left(\frac{i}{i+1}\right)^2 p + \left(\frac{1}{i+1}\right)^2 \eta \|\theta\|^2.$$

Note, under the prior $\pi_i(\theta \,|\, \eta) \pi(\eta)$, we have

$$\iint \eta \|\theta\|^2 \pi_i(\theta \,|\, \eta) \pi(\eta) d\theta d\eta = ip$$

and hence the Bayes risk under $\pi_i(\theta)$ is

$$r(\hat{\theta}_i, \pi_i) = \iint R(\hat{\theta}_i; \theta, \eta)\pi_i(\theta \mid \eta)\pi(\eta)d\theta d\eta = \left(\frac{i}{i+1}\right)p.$$

Therefore, for any estimator δ, we have

$$\sup_{\theta, \eta} R(\delta; \theta, \eta) \ge r(\delta, \pi_i) \ge r(\hat{\theta}_i, \pi_i) = \left(\frac{i}{i+1}\right)p,$$

for all $i \in \mathbb{N}$ and hence

$$\sup_{\theta, \eta} R(\delta; \theta, \eta) \ge p = R(X; \theta, \eta) = \sup_{\theta, \eta} R(X; \theta, \eta),$$

which completes the proof of minimaxity of X. $\qquad\square$

As in Theorem 1.4, we have the following admissibility result for $p = 1, 2$.

Theorem 1.7 *The estimator X is admissible for $p = 1, 2$.*

Proof Recall the joint probability density of X and S is

$$\phi(x - \theta; \eta)f_n(s; \eta) = \frac{\eta^{p/2}}{(2\pi)^{p/2}}\exp\left(-\frac{\eta\|x - \theta\|^2}{2}\right)\frac{\eta^{n/2}s^{n/2-1}}{\Gamma(n/2)2^{n/2}}\exp\left(-\frac{\eta s}{2}\right).$$

The Bayes estimator with respect to the density $\pi(\eta)\eta^{p/2}k_i^2(\eta\|\theta\|^2)$, where k_i is given by (1.32) and $\pi(\eta)$ is any proper prior, is

$$
\begin{aligned}
\hat{\theta}_i(x, s) &= \frac{\iint \eta\theta\phi(x - \theta; \eta)f_n(s; \eta)\pi(\eta)\eta^{p/2}k_i^2(\eta\|\theta\|^2)d\theta d\eta}{\iint \eta\phi(x - \theta; \eta)f_n(s; \eta)\pi(\eta)\eta^{p/2}k_i^2(\eta\|\theta\|^2)d\theta d\eta} \\
&= x + \frac{\iint \eta(\theta - x)\phi(x - \theta; \eta)f_n(s; \eta)\pi(\eta)\eta^{p/2}k_i^2(\eta\|\theta\|^2)d\theta d\eta}{\iint \eta\phi(x - \theta; \eta)f_n(s; \eta)\pi(\eta)\eta^{p/2}k_i^2(\eta\|\theta\|^2)d\theta d\eta} \\
&= x + \frac{\iint \phi(x - \theta; \eta)f_n(s; \eta)\pi(\eta)\eta^{p/2}\nabla_\theta k_i^2(\eta\|\theta\|^2)d\theta d\eta}{\iint \eta\phi(x - \theta; \eta)f_n(s; \eta)\pi(\eta)\eta^{p/2}k_i^2(\eta\|\theta\|^2)d\theta d\eta},
\end{aligned}
\tag{1.43}
$$

where the last equality follows from Lemma 1.1 and (1.28). Note the identity

$$\|x - \theta\|^2 - \|\hat{\theta}_i - \theta\|^2 = -\|\hat{\theta}_i - x\|^2 + 2(\hat{\theta}_i - x)^{\mathsf{T}}(\theta - x). \tag{1.44}$$

Then, by (1.43) and (1.44), the non-scaled Bayes risk difference is

$$\Delta_i = \iint \left(E[\eta \|X - \theta\|^2] - E[\eta \|\hat{\theta}_i(X, S) - \theta\|^2] \right) \pi(\eta)\eta^{p/2}k_i^2(\eta\|\theta\|^2)d\theta d\eta$$

$$= \iint \frac{\| \iint \phi(x - \theta; \eta) f_n(s; \eta)\pi(\eta)\eta^{p/2}\nabla_\theta k_i^2(\eta\|\theta\|^2)d\theta d\eta \|^2}{\iint \eta\phi(x - \theta; \eta) f_n(s; \eta)\pi(\eta)\eta^{p/2}k_i^2(\eta\|\theta\|^2)d\theta d\eta} dxds$$

$$= \iint \frac{\|4 \iint \theta\phi(x - \theta; \eta) f_n(s; \eta)\pi(\eta)\eta^{p/2}\eta k_i(\eta\|\theta\|^2)k_i'(\eta\|\theta\|^2)d\theta d\eta \|^2}{\iint \eta\phi(x - \theta; \eta) f_n(s; \eta)\pi(\eta)\eta^{p/2}k_i^2(\eta\|\theta\|^2)d\theta d\eta} dxds.$$

By the Cauchy-Schwarz inequality (Part 2 of Lemma A.3),

$$\Delta_i \le 16 \iiiint \phi(x - \theta; \eta) f_n(s; \eta)\pi(\eta)\eta^{p/2}\eta\|\theta\|^2\{k_i'(\eta\|\theta\|^2)\}^2 d\theta d\eta dxds$$

$$= 16 \iint \phi(x - \theta; \eta) f_n(s; \eta)dxds \iint \pi(\eta)\eta^{p/2}\eta\|\theta\|^2\{k_i'(\eta\|\theta\|^2)\}^2 d\theta d\eta$$

$$= 16 \frac{\pi^{p/2}}{\Gamma(p/2)} \int_0^\infty \pi(\eta)d\eta \int_0^\infty g^{p/2}\{k_i'(g)\}^2 dg$$

$$\le 16 \frac{\pi^{p/2}}{\Gamma(p/2)} \int_0^\infty (g + 1)\{k_i'(g)\}^2 dg.$$

It follows, as in (1.33) in the proof of Theorem 1.4, that $\Delta_i \to 0$ as $i \to \infty$, which completes the proof.   □

We next address the issue of inadmissibility of X when $p \ge 3$. The risk function of an estimator of the form $\delta_\psi(x, s) = \{1 - \psi(\|x\|^2/s)\}x$ is

$$R(\delta_\psi; \theta, \eta) = E\left[\eta\|\delta_\psi(X, S) - \theta\|^2\right] \tag{1.45}$$

$$= E\left[\eta\|X - \theta\|^2\right] + E\left[\eta\|X\|^2\psi^2(\|X\|^2/S)\right] - 2\sum_{i=1}^p E\left[\eta(X_i - \theta)X_i\psi(\|X\|^2/S)\right].$$

For the second term on the right hand side of (1.45), Lemma 1.2 gives

$$E\left[\eta\|X\|^2\psi^2(\|X\|^2/S)\right] = E\left[\eta S\left(\frac{\|X\|^2}{S}\psi^2(\|X\|^2/S)\right)\right] \tag{1.46}$$

$$= E\left[n\left(\frac{\|X\|^2}{S}\psi^2(\|X\|^2/S)\right) + 2S\left\{-\frac{\|X\|^2}{S^2}\right\}\right.$$

$$\left. \times \left\{\psi^2(\|X\|^2/S) + 2\frac{\|X\|^2}{S}\psi(\|X\|^2/S)\psi'(\|X\|^2/S)\right\}\right].$$

For the third term of the right hand side of (1.45), Lemma 1.1 implies

$$\sum_{i=1}^{p} E\left[\eta(X_i - \theta)X_i\psi\left(\frac{\|X\|^2}{S}\right)\right] = E\left[p\psi\left(\frac{\|X\|^2}{S}\right) + 2\frac{\|X\|^2}{S}\psi'\left(\frac{\|X\|^2}{S}\right)\right]. \quad (1.47)$$

By (1.45), (1.46), and (1.47), we have $R(\delta_\psi; \theta, \eta) = E\left[\hat{R}_\psi(\|X\|^2/S)\right]$, where $\hat{R}_\psi(w)$ is called the SURE (Stein Unbiased Risk Estimate) and is given by

$$\hat{R}_\psi(w) = p + (n-2)w\psi^2(w) - 4w^2\psi(w)\psi'(w) - 2p\psi(w) - 4w\psi'(w).$$
$$(1.48)$$

Let $\psi_{JS}(w) = (p-2)/\{(n+2)w\}$ for $p \geq 3$, which corresponds to the James and Stein (1961) estimator

$$\hat{\theta}_{JS} = \left(1 - \frac{p-2}{n+2}\frac{S}{\|X\|^2}\right)X.$$

Then,

$$\hat{R}_{JS}(w) = p - \frac{(p-2)^2}{n+2}\frac{1}{w},$$

which implies that the risk of the estimator $\hat{\theta}_{JS}$ is smaller than p, as

$$R(\hat{\theta}_{JS}; \theta, \eta) = E[\hat{R}_{JS}(\|X\|^2/S)] = E\left[p - \frac{(p-2)^2}{n+2}\frac{S}{\|X\|^2}\right]$$
$$\leq p = R(X; \theta, \eta). \quad (1.49)$$

Hence we have the following result.

Theorem 1.8 *The estimator X is inadmissible for $p \geq 3$.*

Further the James–Stein estimator is inadmissible since it is dominated by the James–Stein positive-part estimator

$$\hat{\theta}_{JS}^+ = \max\left(0, 1 - \frac{p-2}{n+2}\frac{S}{\|X\|^2}\right)X.$$

In Sect. 3.8, we consider improvements of the James–Stein estimator.

References

Baranchik AJ (1964) Multiple regression and estimation of the mean of a multivariate normal distribution. Technical Report 51, Department of Statistics, Stanford University

Blyth CR (1951) On minimax statistical decision procedures and their admissibility. Ann Math Statist 22:22–42

Brandwein AC, Strawderman WE (1990) Stein estimation: the spherically symmetric case. Statist
    Sci 5(3):356–369
Brown LD, Hwang JT (1982) A unified admissibility proof. In: Statistical decision theory and
    related topics, III, vol 1 (West Lafayette, Ind., 1981). Academic, New York, pp 205–230
Efron B, Morris C (1972a) Empirical Bayes on vector observations: an extension of Stein's method.
    Biometrika 59:335–347
Efron B, Morris C (1972b) Limiting the risk of Bayes and empirical Bayes estimators. II. The
    empirical Bayes case. J Am Statist Assoc 67:130–139
Efron B, Morris C (1976) Families of minimax estimators of the mean of a multivariate normal
    distribution. Ann Statist 4(1):11–21
Fourdrinier D, Strawderman WE, Wells MT (2018) Shrinkage estimation. Springer series in statis-
    tics, Springer, Cham
James W, Stein C (1961) Estimation with quadratic loss. In: Proceedings of the 4th berkeley sym-
    posium on mathematical statistics and probability, vol I. University of California Press, Berkeley,
    California, pp 361–379
Kubokawa T (1991) An approach to improving the James–Stein estimator. J Multivariate Anal
    36(1):121–126
Maruyama Y (2009) An admissibility proof using an adaptive sequence of smoother proper priors
    approaching the target improper prior. J Multivariate Anal 100(8):1845–1853
Maruyama Y, Takemura A (2008) Admissibility and minimaxity of generalized Bayes estimators
    for spherically symmetric family. J Multivariate Anal 99(1):50–73
Stein C (1956) Inadmissibility of the usual estimator for the mean of a multivariate normal distribu-
    tion. In: Proceedings of the third berkeley symposium on mathematical statistics and probability,
    1954–1955, vol I. University of California Press, Berkeley and Los Angeles, pp 197–206
Stein C (1974) Estimation of the mean of a multivariate normal distribution. In: Proceedings of the
    prague symposium on asymptotic statistics, vol II. Charles University, Prague, 1973, pp 345–381
Stein C (1962) Confidence sets for the mean of a multivariate normal distribution. J R Statist Soc
    Ser B 24:265–296
Strawderman WE (1971) Proper Bayes minimax estimators of the multivariate normal mean. Ann
    Math Statist 42(1):385–388

Chapter 2
Estimation of a Normal Mean Vector Under Known Scale

2.1 Introduction

In this chapter, we consider estimation of the mean of a multivariate normal distribution when the scale is known. Without essential loss of generality, we take $\eta = 1$ in this chapter unless otherwise specified. For estimation of μ, the loss function is quadratic loss $L(\delta; \theta, \eta) = \|\delta - \mu\|^2$.

Let $\Pi(\mathrm{d}\mu)$ and m_Π be the prior measure and the corresponding marginal density given by

$$m_\Pi(x) = \int_{\mathbb{R}^p} \phi(x - \mu) \Pi(\mathrm{d}\mu). \tag{2.1}$$

Then the (generalized) Bayes estimator under $\Pi(\mathrm{d}\mu)$ is given by

$$\hat{\mu}_\Pi = \frac{\int_{\mathbb{R}^p} \mu \phi(x - \mu) \Pi(\mathrm{d}\mu)}{\int_{\mathbb{R}^p} \phi(x - \mu) \Pi(\mathrm{d}\mu)} = x + \frac{\int_{\mathbb{R}^p} (\mu - x) \phi(x - \mu) \Pi(\mathrm{d}\mu)}{\int_{\mathbb{R}^p} \phi(x - \mu) \Pi(\mathrm{d}\mu)}$$

$$= x + \frac{\nabla_x m_\Pi(x)}{m_\Pi(x)} = x + \nabla_x \log m_\Pi(x). \tag{2.2}$$

The identity in the second line is known as Brown (1971) identity or Tweedie's formula (Efron 2011).

In Sect. 2.3, we present a sufficient condition for inadmissibility of generalized Bayes estimators under this general prior $\Pi(\mathrm{d}\mu)$.

When considering admissibility in Sect. 2.4, we assume the prior density is given by

$$\pi(\mu) = \int_0^\infty \frac{g^{-p/2}}{(2\pi)^{p/2}} \exp\left(-\frac{\|\mu\|^2}{2g}\right) \Pi(\mathrm{d}g), \tag{2.3}$$

Y. Maruyama et al., *Stein Estimation*, JSS Research Series in Statistics, https://doi.org/10.1007/978-981-99-6077-4_2

for a non-negative measure Π on g. We determine admissibility/inadmissibility of generalized Bayes estimators for a certain subclass of mixture priors in terms of Π. The proof is self-contained allowing us to avoid the complexity of the deeper and more general development in Brown (1971). If Π is finite or proper, the corresponding Bayes estimator is admissible. While we are mainly interested in the case of infinite Π, we do not exclude the case of a finite measure Π.

By Lemma A.1, the marginal likelihood m_π is

$$
\begin{aligned}
m_\pi(\|x\|^2) &= \int_{\mathbb{R}^p} \phi(x - \mu)\pi(\mu)d\mu \\
&= \int_0^\infty \frac{1}{(2\pi)^{p/2}} \exp\left(-\frac{\|x\|^2}{2(g+1)}\right) \frac{\Pi(dg)}{(g+1)^{p/2}},
\end{aligned}
\tag{2.4}
$$

which is finite for all x if

$$
\int_0^\infty \frac{\Pi(dg)}{(g+1)^{p/2}} < \infty.
\tag{2.5}
$$

Throughout this chapter, we assume the prior Π satisfies (2.5). By (2.2), together with m_π given in (2.4), the generalized Bayes estimator under $\pi(\mu)$ given by (2.3) is written as

$$
\begin{aligned}
\hat{\mu}_\pi &= x + \nabla_x \log m_\pi(\|x\|^2) \\
&= \left(1 - \frac{\int_0^\infty (g+1)^{-p/2-1} \exp(-\|x\|^2/\{2(g+1)\})\Pi(dg)}{\int_0^\infty (g+1)^{-p/2} \exp(-\|x\|^2/\{2(g+1)\})\Pi(dg)}\right)x.
\end{aligned}
\tag{2.6}
$$

We give a sufficient condition for admissibility of (generalized) Bayes estimators (2.6) under the prior (2.3) in Sect. 2.4.2.

In order to explore the boundary between admissibility and inadmissibility, $\Pi(dg)$ in (2.3) is assumed to have a regularly varying density of the form

$$
\pi(g; a, b, c) = \frac{1}{(g+1)^a}\left(\frac{g}{g+1}\right)^b \frac{1}{\{\log(g+1)+1\}^c},
\tag{2.7}
$$

$$
\text{for } a > -p/2+1, \ b > -1, \ c \in \mathbb{R},
\tag{2.8}
$$

where (2.8) is sufficient for (2.5). We provide a sufficient condition for admissibility of generalized Bayes estimators (2.6) under the prior (2.7) in Sect. 2.4.3.

Remark 2.1 The term $1/(g+1)^a$ in (2.7) comes from Strawderman (1971) and Berger (1976). The term $\{g/(g+1)\}^b$ in (2.7) comes from Alam (1973) ($-1 < b < 0$) and Faith (1978) ($b \geq 0$). The term $1/\{\log(g+1)+1\}^c$ comes from Maruyama and Strawderman (2020).

When minimaxity is considered in Sect. 2.5, we assume that $\Pi(dg)$ in (2.3) has a regularly varying density $\pi(g) = (g + 1)^{-a}\xi(g)$, where $\xi(g)$ satisfies

AS.1 $\xi(g)$ is continuously differentiable.

AS.2 $\xi(g)$ is slowly varying with $\lim\limits_{g \to \infty} g\dfrac{\xi'(g)}{\xi(g)} = 0$.

We provide a sufficient condition for minimaxity of (generalized) Bayes estimators. Sect. 2.6 considers improvements over the James–Stein estimator.

2.2 Review of Admissibility/Inadmissibility Results

We are interested in determining admissibility/inadmissibility of generalized Bayes estimators, for which Brown (1971) has given an essentially complete solution. Among the many results of Brown (1971), the following result seems to be the most often quoted.

Theorem 2.1 (Theorem 6.3.1 of Brown 1971) *Suppose Π (and hence also $m_\Pi(x)$) is spherically symmetric, and let $m_\Pi(\|x\|^2) := m_\Pi(x)$. If*

$$\int_1^\infty \frac{dt}{t^{p/2}m_\Pi(t)} < \infty, \tag{2.9}$$

then $\hat{\mu}_\Pi$ is inadmissible. If the integral is infinite and the risk of $\hat{\mu}_\Pi$ is bounded or equivalently

$$\sup_\mu E[\|\nabla_x \log m_\Pi(\|X\|^2)\|^2] < \infty, \tag{2.10}$$

then $\hat{\mu}_\Pi$ is admissible.

Suppose $\Pi(dg)$ in (2.3) is assumed to have a regularly varying density $\pi(g; a, b, c)$ given by (2.7). Then a Tauberian theorem (see, e.g., Theorem 13.5.4 in Feller 1971) gives

$$\lim_{t \to \infty} \frac{t^{p/2-1}m_\pi(t; a, b, c)}{\pi(t; a, b, c)} = \frac{\Gamma(p/2 - 1 + a)2^{p/2-1+a}}{(2\pi)^{p/2}} \tag{2.11}$$

where

$$m_\pi(\|x\|^2; a, b, c) = \int_0^\infty \frac{1}{(2\pi)^{p/2}} \exp\left(-\frac{\|x\|^2}{2(g+1)}\right) \frac{\pi(g; a, b, c)}{(g+1)^{p/2}} dg.$$

Hence the integrability (or non-integrability) of (2.9) is equivalent to integrability (or non-integrability) of

$$I(a, b, c) = \int_1^\infty \frac{dg}{g\pi(g; a, b, c)}, \tag{2.12}$$

where, as in Lemma A.8 in Sect. A.3,

$$I(a, b, c) \begin{cases} < \infty & \text{either } a < 0 \text{ or } \{a = 0 \text{ and } c < -1\}, \\ = \infty & \text{either } a > 0 \text{ or } \{a = 0 \text{ and } c \geq -1\}. \end{cases}$$

Further the risk of the corresponding (generalized) Bayes estimator is bounded since

$$\sup_x \| \nabla_x \log m_\pi(\|x\|^2; a, b, c) \|^2 < \infty,$$

which is shown in Lemma A.7 in Sect. A.3. Hence, given the result of Brown (1971), we have the following result.

Theorem 2.2 *Suppose the prior $\Pi(g)$ has a regularly varying density, $\pi(g; a, b, c)$ given by (2.7). Then admissibility/inadmissibility of the the corresponding (generalized) Bayes estimator with $\pi(g; a, b, c)$ is determined by non-integrability/ integrability of $I(a, b, c)$ given by (2.12).*

In this book, we will provide a self-contained proof of Theorem 2.2, based on Maruyama and Strawderman (2023b). Further, our more general sufficient condition for admissibility in Theorem 2.4,

$$\int_0^\infty \frac{\Pi(dg)}{g+1} < \infty$$

includes the case where the risk of the estimator is not bounded. See Remark 2.4.

2.3 Inadmissibility

For the inadmissibility part of Brown's theorem, the following proof is essentially due to Dasgupta and Strawderman (1997), which relates inadmissibility to solving Riccati differential equations. Following Brown (1971) and Dasgupta and Strawderman (1997), we do not assume (2.3) but just spherical symmetry of Π. Hence the result is presented in terms of m_Π (2.1) and $\hat{\mu}_\Pi$ (2.2), not m_π (2.4) and $\hat{\mu}_\pi$ (2.6). With

$$-x\psi(x) = \nabla \log m_\Pi(\|x\|^2) = 2x \frac{m'_\Pi(\|x\|^2)}{m_\Pi(\|x\|^2)}$$

in (1.36), the SURE of $\hat{\mu}_\Pi$ given by (2.2), is

$$\hat{R}_\Pi(\|x\|^2) = p - 4w\left(\frac{m'_\Pi(w)}{m_\Pi(w)}\right)^2 + 4p\frac{m'_\Pi(w)}{m_\Pi(w)} + 8w\frac{m''_\Pi(w)}{m_\Pi(w)}, \tag{2.13}$$

where $w = \|x\|^2$. Using (2.13), we have the following result.

Theorem 2.3 *The generalized Bayes estimator $\hat{\mu}_\Pi$ under the spherically symmetric prior $\Pi(d\mu)$ given by (2.2), is inadmissible if*

$$\int_1^\infty \frac{dt}{t^{p/2}m_\Pi(t)} < \infty,$$

where $m_\Pi(\|x\|^2) := m_\Pi(x)$.

Notice that the statement of Theorem 2.3 is equivalent to the inadmissibility condition of Brown (1971). Hence, by (2.11), the inadmissibility part of Theorem 2.2 follows. More concretely, by Lemma A.8, the integrability of $I(a, b, c)$ given by (2.12) corresponds to the case either $a < 0$ or $\{a = 0$ and $c < -1\}$ and hence we have the following corollary.

Corollary 2.1 *The generalized Bayes estimator with mixing density $\pi(g; a, b, c)$ is inadmissible if*

either $\{-p/2 + 1 < a < 0, \ b > -1, \ c \in \mathbb{R}\}$ or $\{a = 0, \ b > -1, \ c < -1\}$.

For these values of a and c, the integral (2.9) and $I(a, b, c)$ given by (2.12) converges.

Proof (Theorem 2.3) As in (2.13), the SURE of the estimator $\hat{\mu}_\Pi$ given by (2.2) is

$$\hat{R}_\Pi(\|x\|^2) = p - 4w\left(\frac{m'_\Pi(w)}{m_\Pi(w)}\right)^2 + 4p\frac{m'_\Pi(w)}{m_\Pi(w)} + 8w\frac{m''_\Pi(w)}{m_\Pi(w)},$$

where $w = \|x\|^2$. Similarly the SURE of the estimator

$$\hat{\mu}_{\Pi,q} = \hat{\mu}_\Pi - 2\frac{q(\|x\|^2)}{m_\Pi(\|x\|^2)}x \tag{2.14}$$

is $\hat{R}_{\Pi,q}(\|x\|^2) = \hat{R}_\Pi(\|x\|^2) + \Delta(\|x\|^2; q)$ where

$$\Delta(w; q) = 4w\frac{q^2(w)}{m_\Pi(w)}\left(\frac{1}{m_\Pi(w)} - \frac{pq(w) + 2wq'(w)}{wq^2(w)}\right).$$

Now let $r(w) = w^{p/2}q(w)$. Then

$$\frac{1}{m_\Pi(w)} - \frac{pq(w) + 2wq'(w)}{wq^2(w)} = 2w^{p/2}\left\{\frac{1}{2w^{p/2}m_\Pi(w)} + \frac{d}{dw}\left(\frac{1}{r(w)}\right)\right\}.$$

Assume

$$\gamma = \int_1^\infty \frac{dt}{t^{p/2} m_\Pi(t)} < \infty. \tag{2.15}$$

Then a solution of the differential equation $\Delta(w; q) = 0$ is given by

$$\frac{1}{r_*(w)} = -\frac{1}{2} \int_1^w \frac{dt}{t^{p/2} m_\Pi(t)} + \gamma.$$

Hence, under (2.15), the estimator

$$\hat{\mu}_{\Pi, q_*} = \hat{\mu}_\Pi - 2 \frac{q_*(\|x\|^2)}{m_\Pi(\|x\|^2)} x$$

with $q_*(w) = w^{-p/2} r_*(w)$ has the same risk as that of $\hat{\mu}_\Pi$. Since quadratic loss is strictly convex, the estimator given by the average of $\hat{\mu}_\Pi$ and $\hat{\mu}_{\Pi, q_*}$,

$$\hat{\mu}_\Pi - \frac{q_*(\|x\|^2)}{m_\Pi(\|x\|^2)} x, \tag{2.16}$$

strictly improves on $\hat{\mu}_\Pi$. $\square$

Remark 2.2 Suppose $\int_1^\infty dt / \{t^{p/2} m_\Pi(t)\} = \infty$ in (2.15). Then a solution of $\Delta(w; q) = 0$ is

$$\frac{1}{r_\sharp(w)} = -\frac{1}{2} \int_1^w \frac{dt}{t^{p/2} m_\Pi(t)} + C$$

where $1/r_\sharp(w)$ is decreasing with

$$\lim_{w \to 0} \frac{1}{r_\sharp(w)} = +\infty \text{ and } \lim_{w \to \infty} \frac{1}{r_\sharp(w)} = -\infty.$$

Hence $1/r_\sharp(w)$ takes the value 0 once. Let $1/r_\sharp(w_*) = 0$. Then we see that the corresponding $q(w)$ satisfies $\lim_{w \to w_*} q(w) = \infty$ and thus the SURE is not defined for such a $q(w)$ in (2.14).

Remark 2.3 Note

$$\lim_{w \to 0} w q_*(w) = \lim_{w \to 0} \frac{w^{-p/2+1}}{-(1/2) \int_1^w dt / \{t^{p/2} m_\Pi(t)\} + \gamma}$$

$$= \lim_{w \to 0} \frac{(-p/2 + 1) w^{-p/2}}{-1 / \{2 w^{p/2} m_\Pi(w)\}} = (p - 2) m_\Pi(0),$$

where the second equality follows from l'Hôpital's rule. Then the shrinkage factor of $\hat{\mu}_\Pi - \{q_*(\|x\|^2)/m_\Pi(\|x\|^2)\}x$, defined by

$$1 + 2\frac{m'_\Pi(\|x\|^2)}{m_\Pi(\|x\|^2)} - \frac{q_*(\|x\|^2)}{m_\Pi(\|x\|^2)},$$

approaches $-\infty$ as $\|x\|^2 \to 0$, which implies that the average estimator (2.16) is dominated by its positive-part estimator. Hence the average estimator improves on $\hat{\mu}_\Pi$, but is still inadmissible.

Generally speaking, for an inadmissible generalized Bayes estimator $\hat{\mu}_\Pi$, it is difficult to find an admissible estimator which dominates $\hat{\mu}_\Pi$. On page 863 of Brown (1971), there is some interesting related discussion on this topic.

2.4 Admissibility

In this section we establish conditions for admissibility of a subclass of generalized Bayes estimators using Blyth's method as presented in Theorem 1.2.

2.4.1 The Bayes Risk Difference

Let a sequence of (non-normalized) proper priors be of the form

$$\pi_i(\mu) = \int_0^\infty \frac{g^{-p/2}}{(2\pi)^{p/2}} \exp\left(-\frac{\|\mu\|^2}{2g}\right) k_i^2(g)\Pi(dg)$$

where $k_i(g)$ satisfies $\int_0^\infty k_i^2(g)\Pi(dg) < \infty$ for any fixed i and $\lim_{i\to\infty} k_i(g) = 1$ for any fixed g. Specific choice of $k_i(g)$ will be given by (2.23) in Sect. 2.4.2 and (2.26) in Sect. 2.4.3.

Under the prior $\pi_i(\mu)$, we have

$$
\begin{aligned}
m_i(\|x\|^2) &= \int_{\mathbb{R}^p} \phi(x - \mu)\pi_i(\mu)d\mu \\
&= \int_0^\infty \frac{1}{(2\pi)^{p/2}} \exp\left(-\frac{\|x\|^2}{2(g+1)}\right) \frac{k_i^2(g)}{(g+1)^{p/2}}\Pi(dg)
\end{aligned}
\tag{2.17}
$$

and

$$\hat{\mu}_i = x + \nabla_x \log m_i(\|x\|^2)$$

$$= \left(1 - \frac{\int_0^\infty (g+1)^{-p/2-1} \exp(-\|x\|^2/\{2(g+1)\}) k_i^2(g) \Pi(dg)}{\int_0^\infty (g+1)^{-p/2} \exp(-\|x\|^2/\{2(g+1)\}) k_i^2(g) \Pi(dg)}\right) x. \quad (2.18)$$

The Bayes risk difference between $\hat{\mu}_\pi$ and $\hat{\mu}_i$, with respect to $\pi_i(\mu)$, is

$$\Delta_i = \int_{\mathbb{R}^p} \left\{ R(\hat{\mu}_\pi; \mu) - R(\hat{\mu}_i; \mu) \right\} \pi_i(\mu) d\mu,$$

which is rewritten as

$$\Delta_i = \int_{\mathbb{R}^p} \int_{\mathbb{R}^p} \left\{ \|\hat{\mu}_\pi - \mu\|^2 - \|\hat{\mu}_i - \mu\|^2 \right\} \phi(x - \mu) \pi_i(\mu) d\mu dx$$

$$= \int_{\mathbb{R}^p} \left\{ (\|\hat{\mu}_\pi\|^2 - \|\hat{\mu}_i\|^2) m_i(\|x\|^2) - 2(\hat{\mu}_\pi - \hat{\mu}_i)^{\mathsf{T}} \int_{\mathbb{R}^p} \mu \phi(x - \mu) \pi_i(\mu) d\mu \right\} dx$$

$$= \int_{\mathbb{R}^p} \|\hat{\mu}_\pi - \hat{\mu}_i\|^2 m_i(\|x\|^2) dx. \quad (2.19)$$

By (2.6), (2.17) and (2.18), the integrand of Δ_i given in (2.19) is

$$\|\hat{\mu}_\pi - \hat{\mu}_i\|^2 m_i(w)$$

$$= w \left(\frac{\int_0^\infty (g+1)^{-1} F(g; w) k_i^2(g) \Pi(dg)}{\int_0^\infty F(g; w) k_i^2(g) \Pi(dg)} - \frac{\int_0^\infty (g+1)^{-1} F(g; w) \Pi(dg)}{\int_0^\infty F(g; w) \Pi(dg)} \right)^2$$

$$\times \int_0^\infty F(g; w) k_i^2(g) \Pi(dg), \quad (2.20)$$

where $w = \|x\|^2$ and

$$F(g; w) = \frac{(g+1)^{-p/2}}{(2\pi)^{p/2}} \exp\left(-\frac{w}{2(g+1)}\right). \quad (2.21)$$

2.4.2   A General Admissibility Result for Mixture Priors

This subsection is devoted to establishing the following result.

Theorem 2.4 (Maruyama and Strawderman 2023b) *The (generalized) Bayes estimator $\hat{\mu}_\pi$ given by (2.6) is admissible if*

$$\int_0^\infty \frac{\Pi(dg)}{g+1} < \infty. \tag{2.22}$$

Clearly any proper prior on g satisfies (2.22). Further, even if the prior is improper, i.e., $\int_0^\infty \Pi(dg) = \infty$, the corresponding generalized Bayes estimator is admissible under (2.22). Further, by Part 3 of Lemma A.8, we have the following corollary.

Corollary 2.2 *The (generalized) Bayes estimator with mixing density $\pi(g; a, b, c)$ given by (2.7) is admissible if*

$$\text{either } a > \max(-p/2 + 1, 0) \text{ or } \{a = 0 \text{ and } c > 1\}.$$

For these values of a and c, the integral (2.9) and $I(a, b, c)$ given by (2.12) both diverge.

Proof (Proof of Theorem 2.4) Let

$$k_i^2(g) = \frac{i}{g+i}, \tag{2.23}$$

which is increasing in i and is such that $\lim_{i \to \infty} k_i(g) = 1$, for any fixed g. The prior $\pi_i(\mu)$ is proper for any fixed i since the mixture distribution is proper, because

$$\int_0^\infty k_i^2(g)\Pi(dg) \le i \int_0^\infty \frac{\Pi(dg)}{g+1} < \infty.$$

Applying the Cauchy-Schwarz inequality (Part 3 of Lemma A.3), to $\|\hat{\mu}_\pi - \hat{\mu}_i\|^2 m_i(w)$ given by (2.20), we have

$$\|\hat{\mu}_\pi - \hat{\mu}_i\|^2 m_i(w)$$
$$\le 2w \left(\frac{\{\int_0^\infty (g+1)^{-1} F(g; w) k_i^2(g)\Pi(dg)\}^2}{\int_0^\infty F(g; w) k_i^2(g)\Pi(dg)} + \frac{\{\int_0^\infty (g+1)^{-1} F(g; w)\Pi(dg)\}^2}{\int_0^\infty F(g; w)\Pi(dg)} \right), \tag{2.24}$$

where $F(g; w)$ is given by (2.21). Further applying the Cauchy–Schwarz inequality (Part 1 Lemma A.3) to the first and second terms of (2.24), we have

$$\|\hat{\mu}_\pi - \hat{\mu}_i\|^2 m_i(w) \le 4w \int_0^\infty \frac{F(g; w)}{(g+1)^2} \Pi(dg).$$

This is precisely the bound required in order to apply the dominated convergence theorem to demonstrate that $\Delta_i \to 0$, since

$$\lim_{i \to \infty} \hat{\mu}_i = \hat{\mu}_\pi \text{ and hence } \lim_{i \to \infty} \|\hat{\mu}_\pi - \hat{\mu}_i\|^2 = 0 \text{ for fixed } w,$$

and

$$\int_{\mathbb{R}^p} \|\hat{\mu}_\pi - \hat{\mu}_i\|^2 m_i(\|x\|^2) dx$$

$$\leq 4 \int_{\mathbb{R}^p} \int_0^\infty \frac{(2\pi)^{-p/2} \|x\|^2}{(g+1)^{p/2+2}} \exp\left(-\frac{\|x\|^2}{2(g+1)}\right) \Pi(dg) dx$$

$$= \int_{\mathbb{R}^p} \frac{4\|y\|^2 \exp(-\|y\|^2/2) dy}{(2\pi)^{p/2}} \int_0^\infty \frac{\Pi(dg)}{g+1} = 4p \int_0^\infty \frac{\Pi(dg)}{g+1} < \infty,$$

where the second equality follows from Part 2 of Lemma A.2. $\qquad \square$

Remark 2.4 The prior $\mu \sim \mathcal{N}_p(0, gI)$ corresponds to the point prior on g in (2.3). The proper Bayes estimator is $gX/(g+1)$ with unbounded risk

$$\frac{g^2 p + \|\mu\|^2}{(g+1)^2}.$$

Theorem 2.4 covers this case whereas (2.10) of Brown (1971) result is not satisfied by the proper Bayes admissible estimator $gX/(g+1)$.

2.4.3 On the Boundary Between Admissibility and Inadmissibility

By (2.6), the (generalized) Bayes estimator with mixing density $\pi(g; a, b, c)$ is

$$\hat{\mu}_\pi = \left(1 - \frac{\int_0^\infty (g+1)^{-p/2-1} \exp(-\|x\|^2/\{2(g+1)\}) \pi(g; a, b, c) dg}{\int_0^\infty (g+1)^{-p/2} \exp(-\|x\|^2/\{2(g+1)\}) \pi(g; a, b, c) dg}\right) x. \quad (2.25)$$

Corollaries 2.1 and 2.2 settle the issue of admissibility/inadmissibility of $\hat{\mu}_\pi$ given by (2.25) for all values of a and c except for the cases $\{a = 0 \text{ and } |c| \leq 1\}$. For these cases, non-integrability of $I(a, b, c)$ given by (2.12) holds. Hence Theorem 2.2 implies that for these values, near the boundary between admissibility and inadmissibility, the generalized Bayes estimator is admissible.

Theorem 2.5 (Maruyama and Strawderman 2023b) *Assume the measure $\Pi(dg)$ has the density $\pi(g; a, b, c)$ with*

$$a = 0, \ b > -1, \ |c| \leq 1.$$

Then the corresponding generalized Bayes estimator $\hat{\mu}_\pi$ given by (2.25) is admissible.

In Sect. A.4, we prove Theorem 2.5, using the sequence

$$k_i(g) = \begin{cases} 1 - \dfrac{\log(\log(g+1)+1)}{\log(\log(i+1)+1)} & 0 < g < i \\ 0 & g \geq i. \end{cases} \tag{2.26}$$

The proof is based on Maruyama and Strawderman (2023b), where $b \geq 0$ in Theorem 2.5 was assumed. In this book, we include the case $-1 < b < 0$.

2.5 Minimaxity

In this section we study minimaxity of generalized Bayes estimators corresponding to priors of the form $\pi(g) = \xi(g)/(g+1)^a$, where $\xi(g)$ satisfies AS.1 and AS.2 given in the end of Sect. 2.1. The marginal density (2.4) is

$$m_\pi(\|x\|^2) = \int_{\mathbb{R}^p} \phi(x-\mu)\pi(\mu)d\mu$$
$$= \int_0^\infty \frac{1}{(2\pi)^{p/2}} \exp\left(-\frac{\|x\|^2}{2(g+1)}\right) \frac{\xi(g)dg}{(g+1)^{p/2+a}}. \tag{2.27}$$

The SURE of the estimator given by (2.2), is given by

$$\hat{R}(\|x\|^2) = p + 4\frac{m'_\pi(w)}{m_\pi(w)}\left(p - 2w\frac{m''_\pi(w)}{-m'_\pi(w)} + w\frac{-m'_\pi(w)}{m_\pi(w)}\right).$$

Since $m'_\pi(w) \leq 0$, $\hat{R}(w) \leq p$, for all $w \geq 0$, a sufficient condition for minimaxity, is equivalent to $\Delta(v) \geq 0$ for all $v \geq 0$ where $v = w/2$ where

$$\Delta(v) = p - 2\frac{v\int_0^\infty (g+1)^{-p/2-2-a}\xi(g)\exp(-v/(g+1))dg}{\int_0^\infty (g+1)^{-p/2-1-a}\xi(g)\exp(-v/(g+1))dg}$$
$$+ \frac{v\int_0^\infty (g+1)^{-p/2-1-a}\xi(g)\exp(-v/(g+1))dg}{\int_0^\infty (g+1)^{-p/2-a}\xi(g)\exp(-v/(g+1))dg}. \tag{2.28}$$

We consider, separately, the cases where $\xi(g)$ is bounded or unbounded in a neighborhood of the origin in the next two subsections.

2.5.1 $\xi(g)$ *Bounded Near the Origin*

Let

$$\Xi(g) = (g+1)\frac{\xi'(g)}{\xi(g)}.$$

In this subsection, in addition to AS.1 and As.2, we assume the following.

AS.3 $\Xi(g)$ has at most finitely many local extrema for $g \in (0, 1)$.

AS.4 $\displaystyle\lim_{g\to 0} \Xi(g) = \lim_{g\to 0} \frac{\xi'(g)}{\xi(g)} > -\infty.$

Under AS.3, $\Xi(g)$ does not oscillate excessively around the origin. By Part 6 of Lemma A.5 under AS.4, we have $0 \le \xi(0) < \infty$. Thus the assumptions suffice to guarantee that ξ is bounded in a neighborhood of the origin.

Let

$$\Xi_1(g) := \sup_{t \ge g}\left\{(t+1)\frac{\xi'(t)}{\xi(t)}\right\}, \quad \Xi_2(g) := \Xi_1(g) - \Xi(g). \qquad (2.29)$$

Then we have the following result.

Lemma 2.1 *1. $\Xi_1(g)$ is monotone non-increasing and $\displaystyle\lim_{g\to\infty} \Xi_1(g) = 0$.*

2. $\Xi_2(g) \ge 0$ for all $g \ge 0$, $\displaystyle\lim_{g\to\infty} \Xi_2(g) = 0$ and there exists Ξ_{2*} such that

$$0 \le \Xi_2(g) \le \Xi_{2*} \text{ for all } g \ge 0. \qquad (2.30)$$

Proof By definition, $\Xi_1(g)$ is monotone non-increasing and $\Xi_2(g) \ge 0$ for all $g \ge 0$. By AS.2, $\limsup_{g\to\infty} \Xi(g) = \lim_{g\to\infty} \Xi_1(g) = 0$. Further

$$\lim_{g\to\infty} \Xi_2(g) = 0 \qquad (2.31)$$

follows.

By AS.3, there exists g_0 such that $\Xi(g)$ is monotone for $g \in (0, g_0)$. If it is monotone non-increasing, $\Xi_1(g) = \Xi(g)$ and

$$\Xi_2(g) \equiv 0 \text{ for } g \in (0, g_0). \qquad (2.32)$$

If it is monotone non-decreasing, $\Xi_1(g) = \Xi(g_0)$ for $g \in (0, g_0)$ and hence

$$\begin{aligned} \Xi_2(g) &= \Xi_1(g) - \Xi(g) = \Xi(g_0) - \Xi(g) \\ &\le \Xi(g_0) - \lim_{g\to 0} \Xi(g) < \infty \text{ for } g \in (0, g_0), \end{aligned} \qquad (2.33)$$

where the inequality follows from the monotonicity of $\Xi(g)$ and the boundness follows from AS.4. By AS.1, $\Xi(g)$, $\Xi_1(g)$ and $\Xi_2(g)$ are all continuous. Hence, by (2.31), (2.32) and (2.33), there exists Ξ_{2*} as in (2.30). $\qquad\square$

A version of the following result is given in Fourdrinier et al. (1998).

Theorem 2.6 (Fourdrinier et al. 1998) *Assume AS.1–AS.4 on $\xi(g)$. Then the corresponding (generalized) Bayes estimator is minimax if $-p/2 + 1 < a \leq p/2 - 1 - 2\Xi_{2*}$.*

Proof Note $(d/dg) \exp(-v/(g+1)) = v(g+1)^{-2} \exp(-v/(g+1))$. Then

$$v \int_0^\infty \frac{\xi(g) \exp(-v/(g+1))}{(g+1)^{p/2+2+a}} dg$$

$$= \left[\frac{\xi(g) \exp(-v/(g+1))}{(g+1)^{p/2+a}} \right]_0^\infty + (p/2+a) \int_0^\infty \frac{\xi(g) \exp(-v/(g+1))}{(g+1)^{p/2+1+a}} dg$$

$$- \int_0^\infty \frac{\xi'(g) \exp(-v/(g+1))}{(g+1)^{p/2+a}} dg$$

$$= -\frac{\xi(0)}{\exp(v)} + (p/2+a) \int_0^\infty \frac{q(g;v)}{g+1} dg - \int_0^\infty \frac{\Xi(g)q(g;v)}{g+1} dg, \qquad (2.34)$$

where

$$q(g;v) = \frac{\pi(g) \exp(-v/(g+1))}{(g+1)^{p/2}} = \frac{\xi(g) \exp(-v/(g+1))}{(g+1)^{p/2+a}}.$$

Similarly

$$v \int_0^\infty \frac{\xi(g) \exp(-v/(g+1))}{(g+1)^{p/2+1+a}} dg$$

$$= -\frac{\xi(0)}{\exp(v)} + (p/2-1+a) \int_0^\infty q(g;v) dg - \int_0^\infty \Xi(g)q(g;v) dg. \qquad (2.35)$$

By (2.28), (2.34), and (2.35), we have

$$\Delta(v) = -a + \frac{p}{2} - 1 + \frac{\xi(0)}{\exp(v)} \left(\frac{2}{\int_0^\infty (g+1)^{-1}q(g;v)dg} - \frac{1}{\int_0^\infty q(g;v)dg} \right)$$

$$+ 2 \frac{\int_0^\infty (g+1)^{-1}\Xi(g)q(g;v)dg}{\int_0^\infty (g+1)^{-1}q(g;v)dg} - \frac{\int_0^\infty \Xi(g)q(g;v)dg}{\int_0^\infty q(g;v)dg}. \qquad (2.36)$$

In (2.36), we have

$$\frac{2}{\int_0^\infty (g+1)^{-1}q(g;v)dg} - \frac{1}{\int_0^\infty q(g;v)dg} \geq \frac{1}{\int_0^\infty (g+1)^{-1}q(g;v)dg} \geq 0. \quad (2.37)$$

Then, by (2.29), (2.36) and (2.37), we have

$$\Delta(v) \geq -a + \frac{p}{2} - 1 + \Delta_1(v) - \Delta_2(v)$$

where

$$\Delta_i(v) = 2\frac{\int_0^\infty (g+1)^{-1}\Xi_i(g)q(g;v)dg}{\int_0^\infty (g+1)^{-1}q(g;v)dg} - \frac{\int_0^\infty \Xi_i(g)q(g;v)dg}{\int_0^\infty q(g;v)dg}.$$

Since $\Xi_1(g)$ is monotone non-decreasing, the correlation inequality (Lemma A.4) gives

$$\Delta_1(v) \geq \frac{\int_0^\infty (g+1)^{-1}\Xi_1(g)q(g;v)dg}{\int_0^\infty (g+1)^{-1}q(g;v)dg} \geq 0.$$

Further we have $\Delta_2(v) \leq 2\Xi_{2*}$. Hence we have

$$\Delta(v) \geq -a + \frac{p}{2} - 1 - 2\Xi_{2*} \geq 0,$$

which completes the proof.   □

An interesting example is provided by the mixing density $\xi(g) = \{g/(g+1)\}^b \{\log(g+1)+1\}^{-c}$, for $c \in \mathbb{R}$. In this case, $b \geq 0$ is necessary for AS.4. Here is the result.

Lemma 2.2 *Suppose* $\xi(g) = \{g/(g+1)\}^b \{\log(g+1)+1\}^{-c}$, *for* $b \geq 0$ *and* $c \in \mathbb{R}$. *Then*

$$\Xi_{2*} = \begin{cases} 0 & c \leq 0 \\ c/\{\log(b/c+1)+1\} & c > 0. \end{cases}$$

Proof We have

$$\Xi(g) = \frac{b}{g} - \frac{c}{\log(g+1)+1} \quad \text{and} \quad \frac{d}{dg}\Xi(g) = -\frac{b}{g^2} + \frac{c}{g+1}\frac{1}{\{\log(g+1)+1\}^2}.$$

For $b \geq 0$ and $c \leq 0$, $\Xi(g)$ itself is decreasing in g and we can set

$$\Xi_1(g) := \Xi(g) \quad \text{and} \quad \Xi_2(g) \equiv 0.$$

For $b \geq 0$ and $c > 0$,

$$\frac{d}{dg}\Xi(g) = -\frac{b}{g^2} + \frac{c}{g+1}\frac{1}{\{\log(g+1)+1\}^2}.$$

Let $\alpha = b/c$ and $g_* = \alpha\{1 + \log(\alpha + 1)\}$. Then

$$\frac{d}{dg}\Xi(g) = \frac{c}{g^2}\left(-\alpha + \frac{g}{g+1}\frac{g}{\{1 + \log(g+1)\}^2}\right)$$

which is negative for $g \in (0, g_*)$. Hence let

$$\Xi_3(g) = \begin{cases} \Xi(g) & 0 < g < g_* \\ \Xi(g_*) & g \geq g_*, \end{cases}$$

which is decreasing and $\Xi_1(g) \leq \Xi_3(g)$. Then, for $g \geq g_*$, we have

$$\Xi_2(g) = \Xi_1(g) - \Xi(g) \leq \Xi_3(g) - \Xi(g) = \Xi(g_*) - \Xi(g)$$
$$= c\left(\frac{\alpha}{g_*} - \frac{\alpha}{g} - \frac{1}{\log(g_*+1)+1} + \frac{1}{\log(g+1)+1}\right) \leq \frac{c\alpha}{g_*} = \frac{c}{\log(b/c+1)+1},$$

which completes the proof. $\qquad\qquad\square$

The next corollary follows immediately.

Corollary 2.3 *Suppose $b \geq 0$. Then the corresponding (generalized) Bayes estimator with $\pi(g; a, b, c)$ given by (2.7) is minimax if*

$$-p/2 + 1 < a \leq \begin{cases} p/2 - 1 & c \leq 0 \\ p/2 - 1 - \dfrac{2c}{\log(b/c+1)+1} & c > 0. \end{cases}$$

2.5.2 $\xi(g)$ Unbounded Near the Origin

In this subsection we give an example of a class of priors for which $\xi(g)$ is not bounded in a neighborhood of 0, but for which the generalized Bayes estimator is minimax. Suppose $\xi(g) = \{g/(g+1)\}^b$ for $-1 < b < 0$. Then $\xi'(g)/\xi(g) = b/\{g(g+1)\}$ approaches $-\infty$ as $g \to 0$, which does not satisfy AS.4. Further $\lim_{g\to 0}\xi(g) = \infty$ and $\xi(g)$ is unbounded on $g \in (0, 1)$. For such $\xi(g)$, we have the following result.

Theorem 2.7 (Maruyama 1998, 2001) *Assume $-1 < b < 0$ and $c = 0$ in $\pi(g; a, b, c)$. The corresponding (generalized) Bayes estimator is minimax if*

$$-p/2 + 1 < a < p/2 - 1 \text{ and } -\frac{p/2 - 1 - a}{3p/2 + a - 1} \leq b < 0.$$

Maruyama (1998) proved this theorem by expressing the marginal density (2.27) in terms of the confluent hypergeometric function. Here is a more direct proof.

Proof Let $\mathcal{F}(g; v) = \exp(-v/(g + 1)) - \exp(-v)$. Then

$$\frac{\partial}{\partial g}\mathcal{F}(g; v) = \frac{v}{(g + 1)^2}\exp\left(-\frac{v}{g + 1}\right),$$

$$\lim_{g \to 0}\frac{\mathcal{F}(g; v)}{g} = \lim_{g \to 0}\exp\left(-\frac{v}{g + 1}\right)\frac{1}{g}\left\{1 - \exp\left(-\frac{vg}{g + 1}\right)\right\} = \frac{v}{\exp(v)},$$

and hence

$$\lim_{g \to 0}\mathcal{F}(g; v)\xi(g) = 0. \tag{2.38}$$

Note

$$\frac{d}{dg}\left\{\frac{\xi(g)}{(g + 1)^{p/2+a}}\right\} = \left\{\frac{-(p/2 + a)}{g + 1} + \frac{b}{g(g + 1)}\right\}\frac{\xi(g)}{(g + 1)^{p/2+a}}.$$

Then an integration by parts for (2.28) gives

$$v\int_0^\infty \frac{\xi(g)\exp(-v/(g + 1))}{(g + 1)^{p/2+2+a}}dg$$

$$= \left[\frac{\mathcal{F}(g; v)\xi(g)}{(g + 1)^{p/2+a}}\right]_0^\infty - \int_0^\infty \frac{\mathcal{F}(g; v)\xi(g)}{(g + 1)^{p/2+2+a}}\left\{-\frac{p/2 + a}{g + 1} + \frac{b}{g(g + 1)}\right\}dg$$

$$= (p/2 + a)\int_0^\infty \frac{v(g)dg}{(g + 1)\exp(v/(g + 1))} - \frac{p/2 + a}{\exp(v)}\int_0^\infty \frac{v(g)}{g + 1}dg$$

$$- b\int_0^\infty \frac{\mathcal{F}(g; v)v(g)}{g(g + 1)}dg, \tag{2.39}$$

where $v(g) = \xi(g)/(g + 1)^{p/2+a} = \pi(g)/(g + 1)^{p/2}$. Similarly,

$$v\int_0^\infty \frac{\xi(g)\exp(-v/(g + 1))dg}{(g + 1)^{p/2+1+a}} = (p/2 - 1 + a)\int_0^\infty \frac{v(g)dg}{\exp(v/(g + 1))}$$

$$- \frac{p/2 - 1 + a}{\exp(v)}\int_0^\infty v(g)dg - b\int_0^\infty \mathcal{F}(g; v)\frac{v(g)}{g}dg. \tag{2.40}$$

Then, by (2.39) and (2.40),

$$\Delta(v) = \frac{p}{2} - 1 - a + \frac{\Delta_1(v)}{\exp(v)} + 2b\Delta_2(v) - \frac{b\int_0^\infty g^{-1}\mathcal{F}(g; v)v(g)dg}{\int_0^\infty v(g)\exp(-v/(g + 1))dg}, \tag{2.41}$$

where

$$\Delta_1(v) = \frac{2(p/2+a)\int_0^\infty (g+1)^{-1} v(g) dg}{\int_0^\infty (g+1)^{-1} v(g) \exp(-v/(g+1)) dg} - \frac{(p/2-1+a)\int_0^\infty v(g) dg}{\int_0^\infty v(g) \exp(-v/(g+1)) dg},$$

$$\Delta_2(v) = \frac{\int_0^\infty g^{-1}(g+1)^{-1} \mathcal{F}(g; v) v(g) dg}{\int_0^\infty (g+1)^{-1} v(g) \exp(-v/(g+1)) dg}.$$

Since $(g+1)^{-1}$ and $\exp(-v/(g+1))$ are decreasing and increasing in g, respectively, the correlation inequality (Lemma A.4) gives

$$\frac{\int_0^\infty (g+1)^{-1} v(g) dg}{\int_0^\infty (g+1)^{-1} v(g) \exp(-v/(g+1)) dg} \geq \frac{\int_0^\infty v(g) dg}{\int_0^\infty v(g) \exp(-v/(g+1)) dg},$$

and hence $\Delta_1(v) \geq 0$ in (2.41). Let

$$\mathcal{J}(g; v) = \frac{\exp(vg/(g+1)) - 1}{g/(g+1)} = \sum_{j=1}^\infty \left(\frac{g}{g+1}\right)^{j-1} \frac{v^j}{j!}.$$

Then, $\exp(v)\mathcal{F}(g; v) = \exp(vg/(g+1)) - 1$ and

$$\Delta_2(v) \leq \frac{\int_0^\infty \mathcal{J}(g; v)(g+1)^{-2} v(g) dg}{\int_0^\infty \mathcal{J}(g; v) g(g+1)^{-2} v(g) dg}. \qquad (2.42)$$

Further since $\mathcal{J}(g; v)$ is increasing in g, the correlation inequality (Lemma A.4) gives

$$\Delta_2(v) \leq \frac{\int_0^\infty (g+1)^{-2} v(g) dg}{\int_0^\infty g(g+1)^{-2} v(g) dg} = \frac{B(p/2+1+a, b+1)}{B(p/2+a, b+2)} = \frac{p/2+a}{b+1}. \qquad (2.43)$$

Then, by (2.41), (2.42) and (2.43),

$$\Delta(v) \geq p/2 - 1 - a + 2b \frac{p/2+a}{b+1} = \frac{b(3p/2-1+a) + (p/2-1-a)}{b+1} \geq 0,$$

which completes the proof.   □

2.6 Improvement on the James–Stein Estimator

In this section, we construct a class of estimators improving on the James–Stein estimator. From (1.36) and (1.37), the risk difference of two estimators $\hat{\mu}_{\mathrm{JS}}$ and $\hat{\mu}_\phi = \{1 - \phi(\|x\|^2)/\|x\|^2\}x$ can be expressed as

$$\Delta(\lambda) = R(\hat{\mu}_{JS}; \mu) - R(\hat{\mu}_{\phi}; \mu)$$

$$= - E\left[\frac{\{\phi(W) - (p - 2)\}^2}{W}\right] + 4 E[\phi'(W)], \qquad (2.44)$$

for $W = \|X\|^2$ and $\lambda = \|\mu\|^2$. The development in Kubokawa (1994) is particularly useful for deriving conditions on ϕ which suffice to imply $\Delta(\lambda) \geq 0$.

Theorem 2.8 (Kubokawa 1994) *The James–Stein estimator $\hat{\mu}_{JS}$ is improved on by the shrinkage estimator $\hat{\mu}_{\phi}$ if ϕ satisfies the following conditions. (a) $\phi(w)$ is non-decreasing in w; (b) $\lim_{w \to \infty} \phi(w) = p - 2$ and $\phi(w) \geq \phi_0(w)$ where*

$$\phi_0(w) = w \frac{\int_0^\infty (g + 1)^{-p/2-1} \exp(-w/\{2(g + 1)\}) dg}{\int_0^\infty (g + 1)^{-p/2} \exp(-w/\{2(g + 1)\}) dg}.$$

Proof We evaluate the first term $- E[\{\phi(W) - (p - 2)\}^2 / W]$ in (2.44). It follows from condition (b) that

$$- \{\phi(w) - (p - 2)\}^2 = [\{\phi((g + 1)w) - (p - 2)\}^2]_{g=0}^\infty$$

$$= \int_0^\infty \frac{d}{dg} \{\phi((g + 1)w) - (p - 2)\}^2 dg$$

$$= 2w \int_0^\infty \{\phi((g + 1)w) - (p - 2)\} \phi'((g + 1)w) dg, \qquad (2.45)$$

so that the first term may be written as

$$- E\left[\frac{\{\phi(W) - (p - 2)\}^2}{W}\right]$$

$$= 2 \int_0^\infty \int_0^\infty \{\phi((g + 1)w) - (p - 2)\} \phi'((g + 1)w) dg f_p(w; \lambda) dw$$

$$= 2 \int_0^\infty \int_0^\infty \{\phi(v) - (p - 2)\} \phi'(v) \frac{f_p(v/(g + 1); \lambda)}{g + 1} dg dv, \qquad (2.46)$$

where $f_p(w; \lambda)$ is the probability density function of noncentral chi-square distribution $\chi_p^2(\lambda)$ with noncentrality λ.

Replacing v with w in (2.46) gives

$$\Delta(\lambda)$$

$$= 2 \int_0^\infty \phi'(w) \left\{\{\phi(w) - (p - 2)\} \int_0^\infty \frac{f_p(w/(g + 1); \lambda)}{g + 1} dg + 2 f_p(w; \lambda)\right\} dw$$

$$= 2 \int_0^\infty \phi'(w) \left\{\phi(w) - (p - 2) + 2 F_p(w; \lambda)\right\} \left\{\int_0^\infty \frac{f_p(w/(g + 1); \lambda)}{g + 1} dg\right\} dw,$$

where

$$F_p(w; \lambda) = \frac{f_p(w; \lambda)}{\int_0^\infty (g+1)^{-1} f_p(w/(g+1); \lambda) dg}.$$

Note that $F_p(w; \lambda) \geq F_p(w; 0)$ since

$$\frac{1}{F_p(w; \lambda)} - \frac{1}{F_p(w; 0)}$$

$$= \int_0^\infty \frac{1}{g+1} \left\{ \frac{f_p(w/(g+1); \lambda)}{f_p(w; \lambda)} - \frac{f_p(w/(g+1); 0)}{f_p(w; 0)} \right\} dg$$

$$= \int_0^\infty \frac{1}{g+1} \frac{f(w/(g+1); 0)}{f_p(w; \lambda)} \left\{ \frac{f_p(w/(g+1); \lambda)}{f_p(w/(g+1); 0)} - \frac{f_p(w; \lambda)}{f_p(w; 0)} \right\} dg \leq 0, \quad (2.47)$$

where the inequality follows from the fact that

$$\frac{f_p(w; \lambda)}{f_p(w; 0)} = \sum_{i=0}^\infty \frac{(\lambda/2)^i}{e^{\lambda/2} i!} \frac{\Gamma(p/2) 2^{p/2}}{\Gamma(p/2+i) 2^{p/2+i}} \frac{w^{p/2+i-1} e^{-w/2}}{w^{p/2-1} e^{-w/2}}$$

$$= \sum_{i=0}^\infty \frac{(\lambda/2)^i}{e^{\lambda/2} i!} \frac{\Gamma(p/2)}{\Gamma(p/2+i) 2^i} w^i$$

is increasing in w. Hence $\Delta(\lambda) \geq 0$ if $\phi'(w) \geq 0$ and $\phi(w) \geq \phi_0(w)$, where

$$\phi_0(w) = p - 2 - 2 F_p(w; 0), \quad (2.48)$$

which may be expressed as

$$\phi_0(w) = p - 2 - \frac{2 \exp(-w/2)}{\int_0^\infty (g+1)^{-p/2} \exp(-w/\{2(g+1)\}) dg} \quad (2.49)$$

$$= \frac{(p-2) \int_0^\infty (g+1)^{-p/2} \exp(-w/\{2(g+1)\}) dg - 2 \exp(-w/2)}{\int_0^\infty (g+1)^{-p/2} \exp(-w/\{2(g+1)\}) dg}$$

$$= w \frac{\int_0^\infty (g+1)^{-p/2-1} \exp(-w/\{2(g+1)\}) dg}{\int_0^\infty (g+1)^{-p/2} \exp(-w/\{2(g+1)\}) dg}, \quad (2.50)$$

where the last equality follows from an integration by parts. This completes the proof. $\square$

Note that

$$F_p(w; 0) = \frac{1}{\int_0^\infty (g+1)^{-p/2} \exp(wg/\{2(g+1)\}) dg},$$

which is decreasing in w and approaches 0 as $w \to \infty$. Thus, by (2.48) and (2.49),

$$\phi_0'(w) \geq 0, \quad \lim_{w \to \infty} \phi_0(w) = p - 2.$$

Hence the function $\phi_0(w)$ satisfies conditions (a) and (b) of Theorem 2.8. Comparing $\phi_0(w)$ with (2.25), it is clear that

$$\left(1 - \frac{\phi_0(\|X\|^2)}{\|X\|^2}\right) X$$

can be characterized as the generalized Bayes estimator under $\pi(g; a, b, c)$ with $a = b = c = 0$ or, equivalently, the Stein (1974) prior $\pi_S(\|\mu\|^2) = \|\mu\|^{2-p}$ given by (1.14). It follows from Theorems 2.8 and 2.5 that this is estimator is minimax, and admissible, and also improves on the James–Stein estimator.

It follows from (2.50) that $\phi_0(w) \leq w$. Hence the the truncated function

$$\phi_{JS}^+ = \min(w, \, p - 2)$$

corresponding to the James–Stein positive-part estimator

$$\hat{\mu}_{JS}^+ = \max\left(1 - (p - 2)/\|X\|^2, 0\right) X,$$

also satisfies conditions (a) and (b) of Theorem 2.8, and dominates the James–Stein estimator. See Baranchik (1964) and Lehmann and Casella (1998) for the original proof of the domination.

References

Alam K (1973) A family of admissible minimax estimators of the mean of a multivariate normal distribution. Ann Stat 1:517–525

Baranchik AJ (1964) Multiple regression and estimation of the mean of a multivariate normal distribution. Technical Report 51, Department of Statistics, Stanford University

Berger JO (1976) Admissible minimax estimation of a multivariate normal mean with arbitrary quadratic loss. Ann Stat 4(1):223–226

Brown LD (1971) Admissible estimators, recurrent diffusions, and insoluble boundary value problems. Ann Math Stat 42:855–903

Dasgupta A, Strawderman WE (1997) All estimates with a given risk, Riccati differential equations and a new proof of a theorem of Brown. Ann Stat 25(3):1208–1221

Efron B (2011) Tweedie's formula and selection bias. J Am Stat Assoc 106(496):1602–1614. https://doi.org/10.1198/jasa.2011.tm11181

Faith RE (1978) Minimax Bayes estimators of a multivariate normal mean. J Multivariate Anal 8(3):372–379

Feller W (1971) An introduction to probability theory and its applications, vol II, 2nd edn. Wiley, New York

Fourdrinier D, Strawderman WE, Wells MT (1998) On the construction of Bayes minimax estimators. Ann Stat 26(2):660–671

Kubokawa T (1994) A unified approach to improving equivariant estimators. Ann Stat 22(1):290–299

Lehmann EL, Casella G (1998) Theory of point estimation, 2nd edn. Springer texts in statistics. Springer, New York

Maruyama Y (2001) Corrigendum: A unified and broadened class of admissible minimax estimators of a multivariate normal mean. J Multivariate Anal 64(2):196–205 (1998); J Multivariate Anal 78(1):159–160

Maruyama Y (1998) A unified and broadened class of admissible minimax estimators of a multivariate normal mean. J Multivariate Anal 64(2):196–205

Maruyama Y, Strawderman WE (2020) Admissible Bayes equivariant estimation of location vectors for spherically symmetric distributions with unknown scale. Ann Stat 48(2):1052–1071

Maruyama Y, Strawderman WE (2023) A review of Brown 1971 (in)admissibility results under scale mixtures of Gaussian priors. J Stat Plann Inference 222:78–93

Stein C (1974) Estimation of the mean of a multivariate normal distribution. In: Proceedings of the prague symposium on asymptotic statistics (Charles University, Prague, 1973), vol II. Charles University, Prague, pp 345–381

Strawderman WE (1971) Proper Bayes minimax estimators of the multivariate normal mean. Ann Math Stat 42(1):385–388

Chapter 3
Estimation of a Normal Mean Vector Under Unknown Scale

3.1 Equivariance

In this chapter, we consider estimation of the mean of a multivariate normal distribution when the scale is unknown. Let

$$X \sim \mathcal{N}_p(\theta, I/\eta) \quad \text{and} \quad \eta S \sim \chi_n^2,$$

where θ and η are both unknown. For estimation of θ, the loss function is scaled quadratic loss $L(\delta; \theta, \eta) = \eta \| \delta(x, s) - \theta \|^2$.

The first three sections cover issues of Bayesianity, admissibility and minimaxity among estimators which are both orthogonally and scale equivariant. The remaining sections consider these issues among all estimators.

Consider a group of transformations,

$$X \to \gamma \Gamma X, \quad \theta \to \gamma \Gamma \theta, \quad S \to \gamma^2 S, \quad \eta \to \eta/\gamma^2, \tag{3.1}$$

where $\Gamma \in \mathcal{O}(p)$, the group of $p \times p$ orthogonal matrices, and $\gamma \in \mathbb{R}_+$. Equivariant estimators for this group (3.1) satisfy

$$\hat{\theta}(\gamma \Gamma x, \gamma^2 s) = \gamma \Gamma \hat{\theta}(x, s). \tag{3.2}$$

The following result gives the form of such equivariant estimators.

Theorem 3.1 *Equivariant estimators for the group* (3.1) *are of the form*

$$\hat{\theta}_\psi = \left\{ 1 - \psi(\|X\|^2/S) \right\} X, \quad \text{where} \ \psi : \mathbb{R}_+ \to \mathbb{R}.$$

Proof Let the orthogonal matrix $\Gamma \in \mathcal{O}(p)$ satisfy

$$\Gamma = (x/\|x\| \ z_2 \ \dots \ z_p)^{\mathsf{T}} \quad \text{and} \quad \Gamma x = (\|x\| \ 0 \ \dots \ 0)^{\mathsf{T}} = \|x\| e_1, \tag{3.3}$$

where unit vectors $z_2, \dots, z_p \in \mathbb{R}^p$ satisfy

$$z_i^{\mathsf{T}} z_j = 0 \ \text{for} \ i \neq j, \ z_i^{\mathsf{T}} x = 0 \ \text{for} \ i = 2, \dots, p$$

© The Author(s), under exclusive license to Springer Nature Singapore Pte Ltd. 2023
Y. Maruyama et al., *Stein Estimation*, JSS Research Series in Statistics,
https://doi.org/10.1007/978-981-99-6077-4_3

and $e_1 := (1, 0, \ldots, 0)^\top \in \mathbb{R}^p$. Further let $\gamma = 1/\sqrt{s}$. Then, by (3.2), the equivariant estimator $\hat{\theta}(x, s)$ satisfies

$$\hat{\theta}(x, s) = \frac{1}{\gamma}\Gamma^\top\hat{\theta}(\gamma\Gamma x, \gamma^2 s) = \sqrt{s}\Gamma^\top\hat{\theta}(\{\|x\|/\sqrt{s}\}e_1, 1)$$

$$= \frac{\hat{\theta}_1(\{\|x\|/\sqrt{s}\}e_1, 1)}{\|x\|/\sqrt{s}}x + \sqrt{s}\sum_{i=2}^{p}\hat{\theta}_i\left(\frac{\|x\|}{\sqrt{s}}e_1, 1\right)z_i, \tag{3.4}$$

where $\hat{\theta}_i$ is the ith component of $\hat{\theta}$.

For the orthogonal matrix $\Gamma_1\Gamma$ where $\Gamma_1 = \mathrm{diag}(1, -1, 1, \ldots, 1)$, we have

$$\Gamma_1\Gamma = (x/\|x\| \quad - z_2\ z_3\ \ldots\ z_p)^\top \quad \text{and} \quad \Gamma_1\Gamma x = \|x\|e_1.$$

Hence the estimator (3.4) should be also expressed by

$$\hat{\theta}(x, s) = \frac{1}{\gamma}(\Gamma_1\Gamma)^\top\hat{\theta}(\gamma(\Gamma_1\Gamma)x, \gamma^2 s)$$

$$= \frac{\hat{\theta}_1(\{\|x\|/\sqrt{s}\}e_1, 1)}{\|x\|/\sqrt{s}}x - \sqrt{s}\hat{\theta}_2\left(\frac{\|x\|}{\sqrt{s}}e_1, 1\right)z_2 + \sqrt{s}\sum_{i=3}^{p}\hat{\theta}_i\left(\frac{\|x\|}{\sqrt{s}}e_1, 1\right)z_i. \tag{3.5}$$

By (3.4) and (3.5), $\hat{\theta}_2(\{\|x\|/\sqrt{s}\}e_1, 1) = 0$. Similarly, $\hat{\theta}_i(\{\|x\|/\sqrt{s}\}e_1, 1) = 0$ for $i = 3, \ldots, p$. Therefore, in (3.4), we have

$$\hat{\theta}(x, s) = \frac{\hat{\theta}_1(\{\|x\|/\sqrt{s}\}e_1, 1)}{\|x\|/\sqrt{s}}x,$$

where the coefficient of x is a function of $\|x\|^2/s$. This completes the proof. □

Let $f(t) = \{(2\pi)^{p/2}\Gamma(n/2)2^{n/2}\}^{-1}\exp(-t/2)$. Then the joint probability density of X and S is given by

$$\eta^{p/2+n/2}s^{n/2-1}f(\eta\{\|x - \theta\|^2 + s\})$$

$$= \frac{\eta^{p/2}}{(2\pi)^{p/2}}\exp\left(-\frac{\eta\|x - \theta\|^2}{2}\right) \times \frac{\eta^{n/2}s^{n/2-1}}{\Gamma(n/2)2^{n/2}}\exp(-\eta s/2).$$

Also, the generalized Bayes estimator of θ with respect to a prior of the form

$$Q(\theta, \eta; \nu, q) = \eta^\nu\eta^{p/2}q(\eta\|\theta\|^2) \tag{3.6}$$

for $\nu \in \mathbb{R}$ is given by

$$\hat{\theta}_{q,\nu}(x, s) = \frac{\iint \theta\eta^{(2p+n)/2+\nu+1}f(\eta\{\|x - \theta\|^2 + s\})q(\eta\|\theta\|^2)d\theta d\eta}{\iint \eta^{(2p+n)/2+\nu+1}f(\eta\{\|x - \theta\|^2 + s\})q(\eta\|\theta\|^2)d\theta d\eta}. \tag{3.7}$$

The value of the estimator $\hat{\theta}_{q,v}(x, s)$ evaluated at $x = \gamma\Gamma x$ and $s = \gamma^2 s$ where $\Gamma \in \mathcal{O}(p)$, the group of $p \times p$ orthogonal matrices, and $\gamma \in \mathbb{R}_+$, is given by

$$
\hat{\theta}_{q,v}(\gamma\Gamma x, \gamma^2 s)
$$
$$
= \frac{\iint \theta\eta^{(2p+n)/2+v+1} f(\eta\{\|\gamma\Gamma x - \theta\|^2 + \gamma^2 s\})q(\eta\|\theta\|^2)d\theta d\eta}{\iint \eta^{(2p+n)/2+v+1} f(\eta\{\|\gamma\Gamma x - \theta\|^2 + \gamma^2 s\})q(\eta\|\theta\|^2)d\theta d\eta}. \tag{3.8}
$$

By the change of variables $\theta = \gamma\Gamma\theta_*$ and $\eta_* = \gamma^2\eta$, this may be rewritten as

$$
\hat{\theta}_{q,v}(\gamma\Gamma x, \gamma^2 s) = \gamma\Gamma \frac{\iint \theta_*\eta_*^{(2p+n)/2+v+1} f(\eta_*\{\|x - \theta_*\|^2 + s\})q(\eta_*\|\theta_*\|^2)d\theta_* d\eta_*}{\iint \eta_*^{(2p+n)/2+v+1} f(\eta_*\{\|x - \theta_*\|^2 + s\})q(\eta_*\|\theta_*\|^2)d\theta_* d\eta_*}
$$
$$
= \gamma\Gamma\hat{\theta}_{q,v}(x, s). \tag{3.9}
$$

By (3.2), $\hat{\theta}_{q,v}(x, s)$ is equivariant.

In the next section, we show that the case $v = -1$ is special within this class.

3.2 Proper Bayes Equivariant Estimators

In this section we first show that the risk of an estimator that is equivariant under the group (3.1), depends only on the one dimensional parameter $\lambda = \eta\|\theta\|^2 \in \mathbb{R}_+$. We then consider Bayes estimators among the class of equivariant estimators relative to proper priors on λ. We show that such estimators are admissible among equivariant estimators and are also generalized Bayes estimators relative to $Q(\theta, \eta; v, q)$ with $v = -1$ given by (3.6).

Theorem 3.2 *The risk function of an equivariant estimator for the group* (3.1),

$$
\hat{\theta}_\psi = \left\{1 - \psi(\|X\|^2/S)\right\} X
$$

depends only on $\lambda = \eta\|\theta\|^2 \in \mathbb{R}_+$.

Proof As in (3.3), let the orthogonal matrix Γ be of the form

$$
\Gamma^{\mathsf{T}} = (\theta/\|\theta\| \; z_2 \; \cdots \; z_p)^{\mathsf{T}} \quad \text{and} \quad \Gamma^{\mathsf{T}}\theta = (\|\theta\| \; 0 \; \cdots \; 0)^{\mathsf{T}}. \tag{3.10}
$$

By the change of variables, $y = \eta^{1/2}\Gamma^{\mathsf{T}} x$ and $v = \eta s$, we have

$$R(\hat{\theta}_\psi; \theta, \eta)$$

$$= \iint \eta \left\| \{1 - \psi(\|x\|^2/s)\}x - \theta \right\|^2 s^{n/2-1} \eta^{(p+n)/2} f(\eta\{\|x - \theta\|^2 + s\}) dx ds$$

$$= \iint \left\| \{1 - \psi(\|y\|^2/v)\}\Gamma y - \eta^{1/2}\theta \right\|^2 v^{n/2-1} f(\{\|\Gamma y - \eta^{1/2}\theta\|^2 + v\}) dy dv$$

$$= \iint \left\| \{1 - \psi(\|y\|^2/v)\}y - \Gamma^{\mathsf{T}}\eta^{1/2}\theta \right\|^2 v^{n/2-1} f(\{\|y - \Gamma^{\mathsf{T}}\eta^{1/2}\theta\|^2 + v\}) dy dv$$

$$= \iint \left\{ (\{1 - \psi(\|y\|^2/v)\}y_1 - \eta^{1/2}\|\theta\|)^2 + \{1 - \psi(\|y\|^2/v)\}^2 \sum_{i=2}^{p} y_i^2 \right\}$$

$$\times v^{n/2-1} f\left((y_1 - \eta^{1/2}\|\theta\|)^2 + \sum_{i=2}^{p} y_i^2 + v \right) dy dv,$$

where the last equality follows from (3.10). This completes the proof. □

By Theorem 3.2, the risk function may be expressed as

$$R(\hat{\theta}_\psi; \theta, \eta) = \tilde{R}(\hat{\theta}_\psi; \eta\|\theta\|^2). \tag{3.11}$$

Now assume that $\lambda = \eta\|\theta\|^2 \in \mathbb{R}_+$ has the prior density $\bar{\pi}(\lambda)$, which, in this section, we assume to be proper, that is, $\int_0^\infty \bar{\pi}(\lambda) d\lambda = 1$. For an equivariant estimator $\hat{\theta}_\psi$, we define the Bayes equivariant risk as

$$\tilde{r}(\hat{\theta}_\psi; \bar{\pi}) = \int_0^\infty \tilde{R}(\hat{\theta}_\psi; \lambda) \bar{\pi}(\lambda) d\lambda. \tag{3.12}$$

In this book, the estimator $\hat{\theta}_\psi$ which minimizes $\tilde{r}(\hat{\theta}_\psi; \bar{\pi})$, is called a (relative to $\bar{\pi}(\lambda)$). In the following, let

$$\pi(\lambda) = \frac{\Gamma(p/2)}{\pi^{p/2}} \lambda^{1-p/2} \bar{\pi}(\lambda) \tag{3.13}$$

so that $\pi(\|\mu\|^2)$ is a proper probability density on $\mathbb{R}^p$, that is,

$$\int_{\mathbb{R}^p} \pi(\|\mu\|^2) d\mu = 1. \tag{3.14}$$

Let the Bayes equivariant estimator, which minimizes $\tilde{r}(\hat{\theta}_\psi; \bar{\pi})$, be denoted by $\hat{\theta}_\pi$. Theorem 3.3 below shows that $\hat{\theta}_\pi$ is equivalent to the generalized Bayes estimator of

θ with respect to the joint prior density $\eta^{-1}\eta^{p/2}\pi(\eta\|\theta\|^2)$, and that it is admissible among equivariant estimators.

Theorem 3.3 (Maruyama and Strawderman 2020) *Assume that $\bar{\pi}(\lambda)$ is proper.*

1. *The Bayes equivariant risk, $\tilde{r}(\hat{\theta}_\psi; \bar{\pi})$ given by (3.12) is*

$$\tilde{r}(\hat{\theta}_\psi; \bar{\pi}) = \int_{\mathbb{R}^p} \psi(\|z\|^2)\left\{\psi(\|z\|^2) - 2\left(1 - \frac{z^\mathsf{T} M_2(z, \pi)}{\|z\|^2 M_1(z, \pi)}\right)\right\}$$
$$\times \|z\|^2 M_1(z, \pi)\mathrm{d}z + p,$$

where

$$M_1(z, \pi) = \iint \eta^{(2p+n)/2} f(\eta\{\|z - \theta\|^2 + 1\})\pi(\eta\|\theta\|^2)\mathrm{d}\theta\mathrm{d}\eta,$$
$$M_2(z, \pi) = \iint \theta\eta^{(2p+n)/2} f(\eta\{\|z - \theta\|^2 + 1\})\pi(\eta\|\theta\|^2)\mathrm{d}\theta\mathrm{d}\eta. \tag{3.15}$$

2. *Given $\bar{\pi}(\lambda)$, the minimizer of $\tilde{r}(\hat{\theta}_\psi; \bar{\pi})$ with respect to ψ is*

$$\psi_\pi(\|z\|^2) = \arg\min_\psi \tilde{r}(\hat{\theta}_\psi; \bar{\pi}) = 1 - \frac{z^\mathsf{T} M_2(z, \pi)}{\|z\|^2 M_1(z, \pi)}, \tag{3.16}$$

and the Bayes risk difference under $\bar{\pi}(\lambda)$ is

$$\tilde{r}(\hat{\theta}_\psi; \bar{\pi}) - \tilde{r}(\hat{\theta}_\pi; \bar{\pi})$$
$$= \int_{\mathbb{R}^p} \left\{\psi(\|z\|^2) - \psi_\pi(\|z\|^2)\right\}^2 \|z\|^2 M_1(z, \pi)\mathrm{d}z. \tag{3.17}$$

3. *The Bayes equivariant estimator*

$$\hat{\theta}_\pi = \left\{1 - \psi_\pi(\|X\|^2/S)\right\} X$$

with ψ_π by (3.16), is equivalent to the generalized Bayes estimator of θ with respect to the joint prior density $\eta^{-1}\eta^{p/2}\pi(\eta\|\theta\|^2)$ where $\pi(\lambda)$ is given by (3.13).

4. *The Bayes equivariant estimator $\hat{\theta}_\pi$ is admissible within the class of estimators equivariant under the group (3.1).*

Proof (Parts 1 and 2) The Bayes equivariant risk given by (3.12) is

$$\tilde{r}(\hat{\theta}_\psi; \tilde{\pi}) = \int_{\mathbb{R}^p} \tilde{R}(\hat{\theta}_\psi; \|\mu\|^2)\pi(\|\mu\|^2)d\mu$$

$$= \int_{\mathbb{R}^p} \tilde{R}(\hat{\theta}_\psi; \eta\|\theta\|^2)\eta^{p/2}\pi(\eta\|\theta\|^2)d\theta = \int_{\mathbb{R}^p} R(\hat{\theta}_\psi; \theta, \eta)\eta^{p/2}\pi(\eta\|\theta\|^2)d\theta,$$

where the third equality follows from (3.11). Further, expanding terms, $\tilde{r}(\hat{\theta}_\psi; \tilde{\pi})$ may be expressed as

$$\begin{aligned}
\tilde{r}(\hat{\theta}_\psi; \tilde{\pi}) = &\int_{\mathbb{R}^p} E\left[\eta\|X\|^2\psi^2(\|X\|^2/S)\right]\eta^{p/2}\pi(\eta\|\theta\|^2)d\theta \\
&- 2\int_{\mathbb{R}^p} E\left[\eta\|X\|^2\psi(\|X\|^2/S)\right]\eta^{p/2}\pi(\eta\|\theta\|^2)d\theta \\
&+ 2\int_{\mathbb{R}^p} E\left[\eta\psi(\|X\|^2/S)X^\mathsf{T}\theta\right]\eta^{p/2}\pi(\eta\|\theta\|^2)d\theta \\
&+ \int_{\mathbb{R}^p} E\left[\eta\|X - \theta\|^2\right]\eta^{p/2}\pi(\eta\|\theta\|^2)d\theta.
\end{aligned} \tag{3.18}$$

Note that, by the propriety of the prior given by (3.14), the third term is equal to p, that is,

$$\int_{\mathbb{R}^p} E\left[\eta\|X - \theta\|^2\right]\eta^{p/2}\pi(\eta\|\theta\|^2)d\theta = \int_{\mathbb{R}^p} p\pi(\|\mu\|^2)d\mu = p. \tag{3.19}$$

The first and second terms of (3.18) with $\psi^j(\|x\|^2/s)$ for $j = 2, 1$ respectively, may be rewritten as

$$\int_{\mathbb{R}^p} E\left[\eta\|X\|^2\psi^j(\|X\|^2/S)\right]\eta^{p/2}\pi(\eta\|\theta\|^2)d\theta$$

$$= \iiint \eta\|x\|^2\psi^j(\|x\|^2/s)\eta^{(2p+n)/2}s^{n/2-1}f(\eta\{\|x - \theta\|^2 + s\})\pi(\eta\|\theta\|^2)d\theta dx ds$$

$$= \iiint \eta s\|z\|^2\psi^j(\|z\|^2)\eta^{(2p+n)/2}s^{(p+n)/2-1}f(\eta\{\|\sqrt{s}z - \theta\|^2 + s\})$$

$$\qquad \times \pi(\eta\|\theta\|^2)d\theta dz ds \quad (z = x/\sqrt{s}, \ J = s^{p/2})$$

$$= \iiint \eta s\|z\|^2\psi^j(\|z\|^2)\eta^{(2p+n)/2}s^{(2p+n)/2-1}f(s\eta\{\|z - \theta_*\|^2 + 1\})$$

$$\qquad \times \pi(\eta s\|\theta_*\|^2)d\theta_* dz ds \quad (\theta_* = \theta/\sqrt{s}, \ J = s^{p/2})$$

$$= \iiint \|z\|^2\psi^j(\|z\|^2)\eta_*^{(2p+n)/2}f(\eta_*\{\|z - \theta_*\|^2 + 1\})$$

$$\qquad \times \pi(\eta_*\|\theta_*\|^2)d\theta_* dz d\eta_* \quad (\eta_* = \eta s, \ J = 1/\eta)$$

$$= \int_{\mathbb{R}^p} \|z\|^2 \psi^j(\|z\|^2) M_1(z, \pi) dz, \tag{3.20}$$

where $z = x/\sqrt{s}$, J is the Jacobian, and

$$M_1(z, \pi) = \iint \eta^{(2p+n)/2} f(\eta\{\|z - \theta\|^2 + 1\}) \pi(\eta\|\theta\|^2) d\theta d\eta.$$

Similarly, the third term of (3.18) may be rewritten as

$$\int_{\mathbb{R}^p} E\left[\eta\psi(\|X\|^2/S)X^\mathsf{T}\theta\right] \eta^{p/2} \pi(\eta\|\theta\|^2) d\theta = \int_{\mathbb{R}^p} \psi(\|z\|^2) z^\mathsf{T} M_2(z, \pi) dz, \tag{3.21}$$

where

$$M_2(z, \pi) = \iint \theta \eta^{(2p+n)/2} f(\eta\{\|z - \theta\|^2 + 1\}) \pi(\eta\|\theta\|^2) d\theta d\eta.$$

Hence, by (3.19), (3.20) and (3.21), we have

$$\tilde{r}(\hat{\theta}_\psi; \bar{\pi}) = \int_{\mathbb{R}^p} \left\{ \psi^2(\|z\|^2) \|z\|^2 M_1(z, \pi) \right.$$
$$\left. - 2\psi(\|z\|^2)\{\|z\|^2 M_1(z, \pi) - z^\mathsf{T} M_2(z, \pi)\}\right\} dz + p. \tag{3.22}$$

Then the Bayes equivariant solution, or minimizer of $\tilde{r}(\hat{\theta}_\psi; \bar{\pi})$, is

$$\psi_\pi(\|z\|^2) = \arg\min_\psi \tilde{r}(\hat{\theta}_\psi; \bar{\pi}) = 1 - \frac{z^\mathsf{T} M_2(z, \pi)}{\|z\|^2 M_1(z, \pi)} \tag{3.23}$$

and hence the corresponding Bayes equivariant estimator is

$$\hat{\theta}_\pi = \frac{z^\mathsf{T} M_2(z, \pi)}{\|z\|^2 M_1(z, \pi)} x, \tag{3.24}$$

where $z = x/\sqrt{s}$. Parts 1 and 2 follow from (3.22), (3.23) and (3.24).

[Part 3] Note that for $\Gamma \in \mathcal{O}(p)$, the group of $p \times p$ orthogonal matrices, $M_2(\Gamma z, \pi) = \Gamma M_2(z, \pi)$. Hence, as in (3.8) and (3.9), $M_2(z, q)$ is proportional to z and the length of $M_2(z, q)$ is $z^\mathsf{T} M_2(z, q)/\|z\|$, which implies that

$$M_2(z, \pi) = \frac{z^\mathsf{T} M_2(z, q)}{\|z\|} \frac{z}{\|z\|}. \tag{3.25}$$

By (3.25),

$$\hat{\theta}_\pi = \frac{z^\mathsf{T} M_2(z, \pi)}{\|z\|^2 M_1(z, \pi)} x = \sqrt{s} \frac{zz^\mathsf{T} M_2(z, q)}{\|z\|^2 M_1(z, q)} = \sqrt{s} \frac{M_2(z, \pi)}{M_1(z, \pi)}$$

$$= \sqrt{s} \frac{\iint \theta \eta^{(2p+n)/2} f(\eta\{\|x/\sqrt{s} - \theta\|^2 + 1\}) \pi(\eta\|\theta\|^2) d\theta d\eta}{\iint \eta^{(2p+n)/2} f(\eta\{\|x/\sqrt{s} - \theta\|^2 + 1\}) \pi(\eta\|\theta\|^2) d\theta d\eta}.$$

By the change of variables $\theta_* = \sqrt{s}\theta$ and $\eta_* = \eta/s$, we have

$$\hat{\theta}_\pi = \frac{\iint \theta_* \eta_*^{(2p+n)/2} f(\eta_*\{\|x - \theta_*\|^2 + s\}) \pi(\eta_*\|\theta_*\|^2) d\theta_* d\eta_*}{\iint \eta_*^{(2p+n)/2} f(\eta_*\{\|x - \theta_*\|^2 + s\}) \pi(\eta_*\|\theta_*\|^2) d\theta_* d\eta_*},$$

which is the generalized Bayes estimator of θ with respect to $\eta^{-1}\eta^{p/2}\pi(\eta\|\theta\|^2)$, as in (3.7).

[Part 4] Since the quadratic loss function is strictly convex, the Bayes solution is unique, and hence Part 4 follows.   □

As in (3.9), the generalized Bayes estimator of θ with respect to $Q(\theta, \eta; \nu, \pi)$ for any $\nu \in \mathbb{R}$, given by (3.6), is equivariant under the group (3.1). Part 3 of Theorem 3.3, however, applies only to the special case of

$$\nu = -1. \tag{3.26}$$

This is the main reason that we focus on the case of $\nu = -1$ in this book. It should be noted, however, that Theorem 3.3 implies neither admissibility or inadmissibility of generalized Bayes estimators within the class of equivariant estimators, if $\nu \neq -1$.

3.3 Admissible Bayes Equivariant Estimators Through the Blyth Method

Even if $\bar{\pi}(\lambda)$ on $\mathbb{R}_+$ (and hence $\pi(\|\mu\|^2)$ on $\mathbb{R}^p$) is improper, that is

$$\int_{\mathbb{R}^p} \pi(\|\mu\|^2) d\mu = \int_0^\infty \bar{\pi}(\lambda) d\lambda = \infty,$$

the estimator $\hat{\theta}_\pi$ discussed in the previous section can still be defined if $M_1(z, \pi)$ and $M_2(z, \pi)$ given by (3.15) are both finite. The admissibility of such $\hat{\theta}_\pi$ within the class of equivariant estimators can be investigated through Blyth (1951) method.

3.3.1 A General Admissibility Equivariance Result for Mixture Priors

Suppose

$$\bar{\pi}(\lambda) = \int_0^\infty \frac{\lambda^{p/2-1}g^{-p/2}}{2^{p/2}\Gamma(p/2)} \exp\left(-\frac{\lambda}{2g}\right)\Pi(dg)$$

or equivalently

$$\pi(\|\mu\|^2) = \int_0^\infty \frac{g^{-p/2}}{(2\pi)^{p/2}} \exp\left(-\frac{\|\mu\|^2}{2g}\right)\Pi(dg),\qquad(3.27)$$

where $\int_0^\infty \Pi(dg) = \infty$. Then, for (3.15), we have

$$
\begin{aligned}
M_1(z,\pi) &= \iint \eta^{(2p+n)/2} f(\eta\{\|z-\theta\|^2+1\})\pi(\eta\|\theta\|^2)d\theta d\eta \\
&= \frac{1}{q_1(p,n)} \iiint \eta^{(2p+n)/2} \exp\left(-\frac{\eta\{\|z-\theta\|^2+1\}}{2}\right) \\
&\quad \times \frac{1}{(2\pi)^{p/2}g^{p/2}} \exp\left(-\frac{\eta\|\theta\|^2}{2g}\right)\Pi(dg)d\theta d\eta \\
&= \frac{1}{q_1(p,n)} \iint \frac{\eta^{(p+n)/2}}{(g+1)^{p/2}} \exp\left(-\frac{\eta\{\|z\|^2/(g+1)+1\}}{2}\right)\Pi(dg)d\eta \\
&= \frac{\Gamma((p+n)/2+1)}{q_1(p,n)2^{-(p+n)/2-1}} \int_0^\infty \frac{(g+1)^{-p/2}\Pi(dg)}{\{1+\|z\|^2/(g+1)\}^{(p+n)/2+1}},\qquad(3.28)
\end{aligned}
$$

where the third equality follows from Lemma A.1, and

$$q_1(p,n) = (2\pi)^{p/2}\Gamma(n/2)2^{n/2}.\qquad(3.29)$$

Similarly, for (3.15), we have

$$M_2(z,\pi) = \frac{\Gamma((p+n)/2+1)}{q_1(p,n)2^{-(p+n)/2-1}} \int_0^\infty \frac{gz}{g+1} \frac{(g+1)^{-p/2}\Pi(dg)}{\{1+\|z\|^2/(g+1)\}^{(p+n)/2+1}}.\qquad(3.30)$$

Then, by (3.16), (3.28) and (3.30), the (improper or generalized) Bayes equivariant estimator is

$$
\begin{aligned}
\hat{\theta}_\pi &= \{1-\psi_\pi(\|z\|^2)\}x \\
&= \left(1 - \frac{\int_0^\infty (g+1)^{-p/2-1}\{1+\|z\|^2/(g+1)\}^{-(p+n)/2-1}\Pi(dg)}{\int_0^\infty (g+1)^{-p/2}\{1+\|z\|^2/(g+1)\}^{-(p+n)/2-1}\Pi(dg)}\right)x,\qquad(3.31)
\end{aligned}
$$

where $\|z\|^2 = \|x\|^2/s$. For some $k_i^2(g)$, assume the propriety of $k_i^2(g)\Pi(dg)$ as $\int_0^\infty k_i^2(g)\Pi(dg) < \infty$. Then

$$\bar{\pi}_i(\lambda) = \int_0^\infty \frac{\lambda^{p/2-1}}{g^{p/2}2^{p/2}\Gamma(p/2)}\exp\left(-\frac{\lambda}{2g}\right)k_i^2(g)\Pi(dg) \tag{3.32}$$

is also proper. Let $\hat{\theta}_{\pi i} = \{1 - \psi_{\pi i}(\|x\|^2/s)\}x$ be the proper Bayes equivariant estimator under $\bar{\pi}_i(\lambda)$. By (3.17), the Bayes risk difference between $\hat{\theta}_\pi$ and $\hat{\theta}_{\pi i}$ under $\bar{\pi}_i$ is

$$\tilde{r}(\hat{\theta}_\pi; \bar{\pi}_i) - \tilde{r}(\hat{\theta}_{\pi i}; \bar{\pi}_i)$$
$$= \int_{\mathbb{R}^p} \left\{\psi_\pi(\|z\|^2) - \psi_{\pi i}(\|z\|^2)\right\}^2 \|z\|^2 M_1(z, \pi_i)dz. \tag{3.33}$$

For $w = \|z\|^2$, the integrand of (3.33) is expressed as

$$\{\psi_\pi(\|z\|^2) - \psi_{\pi i}(\|z\|^2)\}^2\|z\|^2 M_1(z, \pi_i)$$
$$= w\left(\frac{\int_0^\infty (g+1)^{-p/2-1}\{1 + w/(g+1)\}^{-(p+n)/2-1}\Pi(dg)}{\int_0^\infty (g+1)^{-p/2}\{1 + w/(g+1)\}^{-(p+n)/2-1}\Pi(dg)}\right.$$
$$\left. - \frac{\int_0^\infty (g+1)^{-p/2-1}\{1 + w/(g+1)\}^{-(p+n)/2-1}k_i^2(g)\Pi(dg)}{\int_0^\infty (g+1)^{-p/2}\{1 + w/(g+1)\}^{-(p+n)/2-1}k_i^2(g)\Pi(dg)}\right)^2$$
$$\times \frac{\Gamma((p+n)/2+1)2^{(p+n)/2+1}}{q_1(p,n)}\int_0^\infty \frac{(g+1)^{-p/2}k_i^2(g)\Pi(dg)}{\{1 + w/(g+1)\}^{(p+n)/2+1}}. \tag{3.34}$$

As in Sect. 2.4.2, with the sequence $k_i^2(g) = i/(g+i)$, we have the following result on admissibility within the class of equivariant estimators.

Theorem 3.4 (Maruyama and Strawderman 2020) *The estimator $\hat{\theta}_\pi$ is admissible within the class of equivariant estimators if*

$$\int_0^\infty \frac{\Pi(dg)}{g+1} < \infty.$$

Proof Under the above assumption, $k_i^2(g) = i/(g+i)$ gives an increasing sequence of proper priors since

$$\int_0^\infty k_i^2(g)\Pi(dg) = i\int_0^\infty \frac{\Pi(dg)}{g+i} \le i\int_0^\infty \frac{\Pi(dg)}{g+1} < \infty,$$

for fixed i. Applying the inequality (Part 3 of Lemma A.3) to (3.34), we have

$$\frac{q_1(p,n)}{\Gamma((p+n)/2+1)2^{(p+n)/2+1}}\{\psi_\pi(\|z\|^2)-\psi_{\pi i}(\|z\|^2)\}^2\|z\|^2 M_1(z,\pi_i)$$

$$\leq 2w\left(\frac{\{\int_0^\infty(g+1)^{-p/2-1}\{1+w/(g+1)\}^{-(p+n)/2-1}\Pi(dg)\}^2}{\int_0^\infty(g+1)^{-p/2}\{1+w/(g+1)\}^{-(p+n)/2-1}\Pi(dg)}\right.$$

$$\left.+\frac{\{\int_0^\infty(g+1)^{-p/2-1}\{1+w/(g+1)\}^{-(p+n)/2-1}k_i^2(g)\Pi(dg)\}^2}{\int_0^\infty(g+1)^{-p/2}\{1+w/(g+1)\}^{-(p+n)/2-1}k_i^2(g)\Pi(dg)}\right),$$

where $q_1(p,n)$ is given by (3.29). Further, applying the Cauchy-Schwarz inequality (Part 1 of Lemma A.3) to the first and second terms, we have

$$\frac{q_1(p,n)}{\Gamma((p+n)/2+1)2^{(p+n)/2+1}}\{\psi_\pi(\|z\|^2)-\psi_{\pi i}(\|z\|^2)\}^2\|z\|^2 M_1(z,\pi_i)$$

$$\leq 4\|z\|^2\int_0^\infty\frac{(g+1)^{-p/2-2}\Pi(dg)}{\{1+\|z\|^2/(g+1)\}^{(p+n)/2+1}}.$$

Hence, we have

$$\frac{q_1(p,n)}{\Gamma((p+n)/2+1)2^{(p+n)/2+1}}\int_{\mathbb{R}^p}\{\psi_\pi(\|z\|^2)-\psi_{\pi i}(\|z\|^2)\}^2\|z\|^2 M_1(z,\pi_i)dz$$

$$\leq 4\int_{\mathbb{R}^p}\int_0^\infty\frac{\|z\|^2}{\{1+\|z\|^2/(g+1)\}^{(p+n)/2+1}}\frac{\Pi(dg)}{(g+1)^{p/2+2}}dz$$

$$=4\frac{\pi^{p/2}}{\Gamma(p/2)}\int_0^\infty\int_0^\infty\frac{t^{p/2}}{(1+t)^{(p+n)/2+1}}\frac{\Pi(dg)}{g+1}dt$$

$$=4\frac{\pi^{p/2}}{\Gamma(p/2)}B(p/2+1,n/2)\int_0^\infty\frac{\Pi(dg)}{g+1}<\infty,$$

where the equalities follow from Part 1 of Lemma A.2 and Part 3 of Lemma A.2, respectively.

Then by the dominated convergence theorem, we have

$$\lim_{i\to\infty}\left\{\tilde{r}(\hat{\theta}_\pi;\pi_i)-\tilde{r}(\hat{\theta}_{\pi i};\pi_i)\right\}=0$$

which, by the Blyth method, implies the admissibility of $\hat{\theta}_\pi$ within the class of equivariant estimators. $\qquad\square$

As in Sect. 2.4.3, suppose $\Pi(dg)$ in (3.27) has a regularly varying density of the form

$$\pi(g;a,b,c)=\frac{1}{(g+1)^a}\left(\frac{g}{g+1}\right)^b\frac{1}{\{\log(g+1)+1\}^c}. \qquad (3.35)$$

Then, by (3.31), the corresponding generalized Bayes estimator is of the form

$$\hat{\theta}_\pi = \left(1 - \frac{\int_0^\infty (g+1)^{-p/2-1}\{1+w/(g+1)\}^{-(p+n)/2-1}\pi(g;a,b,c)dg}{\int_0^\infty (g+1)^{-p/2}\{1+w/(g+1)\}^{-(p+n)/2-1}\pi(g;a,b,c)dg}\right)x.$$

(3.36)

As a corollary of Theorem 3.4, using the argument in the admissibility proofs of Sect. 2.4.3, we have the following result.

Corollary 3.1 *The generalized Bayes estimator $\hat{\theta}_\pi$ given by (3.36) is admissible within the class of equivariant estimators if*

$$either \ \{a > 0, \ b > -1, \ c \in \mathbb{R}\} \ or \ \{a = 0, \ b > -1, \ c > 1\}.$$

3.3.2 On the Boundary Between Equivariant Admissibility and Inadmissibility

For the class of densities $\pi(g; a, b, c)$ given by (3.35), with either $-p/2 + 1 < a < 0$ or $\{a = 0$ and $c > 1\}$, Corollaries 3.3 and 3.4 in Sect. 3.6 show the inadmissibility of the corresponding generalized Bayes estimator by finding an improved estimator among the class of equivariant estimators. Hence, together with Corollary 3.1, the issue of admissibility/inadmissibility within the class of equivariant estimators for all values of a and c except for the cases $\{a = 0$ and $|c| \le 1\}$, has been settled. The following result addresses this case.

Theorem 3.5 (Maruyama and Strawderman 2020) *Assume the measure $\Pi(dg)$ in (3.27) has the density $\pi(g; a, b, c)$ given by (3.35) with*

$$a = 0, \ b > -1, \ -1 < c \le 1.$$

Then the corresponding generalized Bayes estimator is admissible within the class of equivariant estimators.

Proof See Appendix A.6.   □

Our proof unfortunately does not cover the case $c = -1$, although we conjecture that admissibility holds within the class of equivariant estimators as well. The proof of Theorem 3.5 is based on Maruyama and Strawderman (2020), where $b \ge 0$ was assumed. In this book, we also include the case $-1 < b < 0$.

While this section considers admissibility only within the class of equivariant estimators, the next section broadens the discussion and considers admissibility among all estimators.

3.4 Admissibility Among All Estimators

3.4.1 The Main Result

In this section, we consider admissibility of generalized Bayes estimators among all estimators for a broad class of mixture priors. In particular, we consider the following class of joint prior densities:

$$\pi_*(\theta, \eta) = \frac{1}{\eta} \times \eta^{p/2} \pi(\eta \|\theta\|^2)$$

where

$$\pi(\|\mu\|^2) = \int_0^\infty \frac{g^{p/2}}{(2\pi)^{p/2}} \exp\left(-\frac{\|\mu\|^2}{2g}\right) \pi(g; a, b, 0)dg, \qquad (3.37)$$

and where $\pi(g; a, b, c)$ is given in (3.35). We note that all such priors are improper because each is non-integrable in η for any given θ. Then, as in (3.36), the corresponding generalized Bayes estimator is

$$\{1 - \phi(\|x\|^2/s)/\{\|x\|^2/s\}\}x$$

where

$$\phi(w) = w \frac{\int_0^\infty (g+1)^{-p/2-1}\{1 + w/(g+1)\}^{-p/2-n/2-1}\pi(g; a, b, 0)dg}{\int_0^\infty (g+1)^{-p/2}\{1 + w/(g+1)\}^{-p/2-n/2-1}\pi(g; a, b, 0)dg}. \qquad (3.38)$$

Here is the main theorem of this section.

Theorem 3.6 (Maruyama and Strawderman 2021, 2023a) *The generalized Bayes estimator under $\pi_*(\theta, \eta)$ is admissible among all estimators if*

$$\max(-p/2 + 1, 0) < a < n/2 + 2, \ b > -1, \ c = 0.$$

Remark 3.1 As far as we know, Theorem 3.6 is the only known result on admissibility of generalized Bayes estimators of the form $\{1 - \phi(\|x\|^2/s)/(\|x\|^2/s)\} x$. As in Corollary 3.5 in Sect. 3.7, the generalized Bayes estimator under $\pi_*(\theta, \eta)$ is also minimax if

$$-p/2 + 1 < a \le \frac{(p-2)(n+2)}{2(2p+n-2)}, \ b \ge 0, \ c = 0.$$

Strawderman (1973) considered the truncated proper prior on η, $\eta^c I_{(\gamma,\infty)}$ with $c < -1$ and $\gamma > 0$ instead of the invariant prior on η. Under this prior, a class of proper Bayes, and hence admissible estimators dominating the usual unbiased estimator for $p \ge 5$ was found. However, because of the truncation of the prior on η, such estimators are not scale equivariant of the form $\left\{1 - \phi(\|x\|^2/s)/(\|x\|^2/s)\right\} x$, but instead have the form $\left\{1 - \phi(\|x\|^2/s, s)/(\|x\|^2/s)\right\} x$.

Recall $\pi(g; a, b, c)$ given by (3.37) is proper for $a > 1$ and $c \in \mathbb{R}$. In order to prove the result, we construct a sequence of proper priors $\pi_i(\theta, \eta)$ converging to $\pi_*(\theta, \eta)$ of the form

$$\pi_i(\theta, \eta) = \frac{h_i^2(\eta)}{\eta} \int_0^\infty \frac{\eta^{p/2}}{(2\pi)^{p/2} g^{p/2}} \exp\left(-\frac{\eta}{2g}\|\theta\|^2\right) \pi(g) k_i^2(g) dg \qquad (3.39)$$

where

$$h_i(\eta) = \frac{\log(i+1)}{\log(i+1) + |\log \eta|},$$

$$k_i(g) = \begin{cases} 1 - \dfrac{\log(g+1)}{\log(g+1+i)} & \max(-p/2+1, 0) < a \le 1, \\ 1 & 1 < a < n/2 + 2. \end{cases}$$

Note that $\log(1+1) < 1 < \log(2+1)$. For this technical reason, the sequence starts at $i = 2$. Properties of $h_i(\eta)$ and $k_i(g)$ are provided in Lemmas 3.1 and A.6. In particular, we emphasize that $h_i^2(\eta)/\eta$ and $\pi(g)k_i^2(g)$ are both proper by Part 2 of Lemma 3.1 and Part 5 of Lemma A.6, respectively, which implies that $\pi_i(\theta, \eta)$ given by (3.39) is proper.

Lemma 3.1 *Let*

$$h_i(\eta) = \frac{\log(i+1)}{\log(i+1) + |\log \eta|}.$$

1. *$h_i(\eta)$ is increasing in i and $\lim_{i \to \infty} h_i(\eta) = 1$ for all $\eta > 0$.*
2. *$\displaystyle \int_0^\infty \eta^{-1} h_i^2(\eta) d\eta = 2\log(i+1).$*

Proof (Part 1) This part is straightforward given the form of $h_i(\eta)$.
　　[Part 2] Let $j = \log(i+1)$. The results follow from the integrals,

$$\int_0^\infty \frac{h_i^2(\eta)}{\eta} d\eta = \int_0^1 \frac{j^2 d\eta}{\eta(j - \log \eta)^2} + \int_1^\infty \frac{j^2 d\eta}{\eta(j + \log \eta)^2}$$

$$= \left[\frac{j^2}{j - \log \eta}\right]_0^1 + \left[\frac{-j^2}{j + \log \eta}\right]_1^\infty = 2j.$$

$\square$

3.4.2 A Proof of Theorem 3.6

We start by developing expressions for Bayes estimators and risk differences which are used to prove Theorem 3.6. We make use of the following notation. For any function $\psi(\theta, \eta)$, let

$$
\begin{aligned}
&m(\psi(\theta, \eta)) \\
&= \iint \psi(\theta, \eta) \frac{\eta^{p/2}}{(2\pi)^{p/2}} \exp\left(-\eta\frac{\|x - \theta\|^2}{2}\right) \frac{\eta^{n/2}s^{n/2-1}}{\Gamma(n/2)2^{n/2}} \exp\left(-\frac{\eta s}{2}\right) d\theta d\eta.
\end{aligned}
$$

Then, under the loss (1.3), the generalized Bayes estimator under the improper prior $\pi_*(\theta, \eta)$ is

$$
\hat{\theta}_* = \frac{m(\eta\theta\pi_*(\theta, \eta))}{m(\eta\pi_*(\theta, \eta))}
$$

and the proper Bayes estimator under the proper prior $\pi_i(\theta, \eta)$ is

$$
\hat{\theta}_i = \frac{m(\eta\theta\pi_i(\theta, \eta))}{m(\eta\pi_i(\theta, \eta))}.
$$

The Bayes risk difference under π_i is

$$
\Delta_i = \int_{\mathbb{R}^p} \int_0^\infty \left\{ \mathrm{E}\left[\eta\|\hat{\theta}_* - \theta\|^2\right] - \mathrm{E}\left[\eta\|\hat{\theta}_i - \theta\|^2\right] \right\} \pi_i(\theta, \eta) d\theta d\eta.
$$

Note that $\|\hat{\theta}_* - \theta\|^2 - \|\hat{\theta}_i - \theta\|^2 = \|\hat{\theta}_*\|^2 - \|\hat{\theta}_i\|^2 - 2\theta^{\mathrm{T}}(\hat{\theta}_* - \hat{\theta}_i)$. Then Δ_i can be re-expressed as

$$
\begin{aligned}
\Delta_i &= \iiiint \eta \left(\|\hat{\theta}_*\|^2 - \|\hat{\theta}_i\|^2 - 2\theta^{\mathrm{T}}(\hat{\theta}_* - \hat{\theta}_i) \right) \\
&\quad \times \frac{\eta^{p/2}}{(2\pi)^{p/2}} \exp\left(-\eta\frac{\|x - \theta\|^2}{2}\right) \frac{\eta^{n/2}s^{n/2-1}}{\Gamma(n/2)2^{n/2}} \exp\left(-\frac{\eta s}{2}\right) \pi_i(\theta, \eta) dx ds d\theta d\eta \\
&= \iint \left\{ m(\eta\pi_i)(\|\hat{\theta}_*\|^2 - \|\hat{\theta}_i\|^2) - 2m(\eta\theta^{\mathrm{T}}\pi_i)(\hat{\theta}_* - \hat{\theta}_i) \right\} dx ds \\
&= \iint \|\hat{\theta}_* - \hat{\theta}_i\|^2 m(\eta\pi_i(\theta, \eta)) dx ds. \quad\quad (3.40)
\end{aligned}
$$

Next, we rewrite $\hat{\theta}_*$, $\hat{\theta}_i$ and $\|\hat{\theta}_* - \hat{\theta}_i\|^2 m(\eta\pi_i(\theta, \eta))$, the integrand of (3.40). By Lemma A.1, we have

$$m(\eta\pi_i) = \iiint \eta \frac{\eta^{p/2}}{(2\pi)^{p/2}} \exp\left(-\eta\frac{\|x-\theta\|^2}{2}\right) \frac{\eta^{n/2}s^{n/2-1}}{\Gamma(n/2)2^{n/2}} \exp\left(-\frac{\eta s}{2}\right)$$

$$\times \frac{\eta^{p/2}}{(2\pi)^{p/2}g^{p/2}} \exp\left(-\frac{\eta}{2g}\|\theta\|^2\right) \frac{h_i^2(\eta)}{\eta} \pi(g)k_i^2(g)d\theta dgd\eta$$

$$= \frac{s^{n/2-1}}{q_1(p,n)} \iint F(g,\eta;w,s)h_i^2(\eta)\pi(g)k_i^2(g)dgd\eta, \tag{3.41}$$

where $w = \|x\|^2/s$, $q_1(p,n) = (2\pi)^{p/2}\Gamma(n/2)2^{n/2}$, and

$$F(g,\eta;w,s) = \frac{\eta^{p/2+n/2}}{(g+1)^{p/2}} \exp\left(-\frac{\eta s}{2}\left(1+\frac{w}{g+1}\right)\right).$$

Similarly we have

$$m(\eta\theta\pi_i) = \frac{s^{n/2-1}}{q_1(p,n)} \iint \frac{gx}{g+1} F(g,\eta;w,s)h_i^2(\eta)\pi(g)k_i^2(g)dgd\eta. \tag{3.42}$$

By (3.41) and (3.42), the Bayes estimator under π_i is

$$\hat{\theta}_i = \frac{m(\theta\eta\pi_i)}{m(\eta\pi_i)} = \left(1 - \frac{\phi_i(w,s)}{w}\right)x, \tag{3.43}$$

where

$$\phi_i(w,s) = w\frac{\iint(g+1)^{-1}F(g,\eta;w,s)h_i^2(\eta)\pi(g)k_i^2(g)dgd\eta}{\iint F(g,\eta;w,s)h_i^2(\eta)\pi(g)k_i^2(g)dgd\eta}. \tag{3.44}$$

With $h_i \equiv 1$ and $k_i \equiv 1$ in (3.44), we have

$$\phi_*(w,s) = w\frac{\iint(g+1)^{-1}F(g,\eta;w,s)\pi(g)dgd\eta}{\iint F(g,\eta;w,s)\pi(g)dgd\eta} \tag{3.45}$$

and our target generalized Bayes estimator given by

$$\hat{\theta}_* = \left(1 - \frac{\phi_*(w,s)}{w}\right)x. \tag{3.46}$$

Note that

$$\int_0^\infty F(g,\eta;w,s)d\eta = \frac{\Gamma(p/2+n/2+1)}{(g+1)^{p/2}}\left(\frac{2s^{-1}}{1+w/(g+1)}\right)^{p/2+n/2+1}$$

which implies

$$\frac{\phi_*(w,s)}{w} = \frac{\int_0^\infty(g+1)^{-p/2-1}\{1+w/(g+1)\}^{-p/2-n/2-1}\pi(g)dg}{\int_0^\infty(g+1)^{-p/2}\{1+w/(g+1)\}^{-p/2-n/2-1}\pi(g)dg}. \tag{3.47}$$

In the following development, however, we utilize (3.45) not (3.47) as the expression of $\phi_*(w, s)$.

By (3.41), (3.43) and (3.46), we have

$$\frac{q_1(p, n)}{\|x\|^2 s^{n/2-1}} \left\| \hat{\theta}_* - \hat{\theta}_i \right\|^2 m(\eta\pi_i) = \frac{q_1(p, n)}{s^{n/2-1}} \left(\frac{\phi_*(w, s)}{w} - \frac{\phi_i(w, s)}{w} \right)^2 m(\eta\pi_i)$$

$$= A(w, s; i),$$

where

$$A(w, s, i) = \left(\frac{\iint \frac{F\pi}{g+1} dgd\eta}{\iint F\pi dgd\eta} - \frac{\iint \frac{Fh_i^2 \pi k_i^2}{g+1} dgd\eta}{\iint Fh_i^2 k_i^2 dgd\eta} \right)^2 \iint Fh_i^2 \pi k_i^2 dgd\eta.$$

(3.48)

Applying the inequality (Part 3 of Lemma A.3) to (3.48), we have

$$\frac{1}{3} \left(\frac{\iint (g+1)^{-1} F\pi dgd\eta}{\iint F\pi dgd\eta} - \frac{\iint (g+1)^{-1} Fh_i^2 \pi k_i^2 dgd\eta}{\iint Fh_i^2 \pi k_i^2 dgd\eta} \right)^2$$

$$\leq \left(\frac{\iint (g+1)^{-1} F\pi dgd\eta}{\iint F\pi dgd\eta} - \frac{\iint (g+1)^{-1} Fh_i^2 \pi dgd\eta}{\iint Fh_i^2 \pi dgd\eta} \right)^2$$

$$+ \left(\frac{\iint (g+1)^{-1} Fh_i^2 \pi dgd\eta}{\iint Fh_i^2 \pi dgd\eta} - \frac{\iint (g+1)^{-1} Fh_i^2 \pi k_i^2 dgd\eta}{\iint Fh_i^2 \pi dgd\eta} \right)^2$$

$$+ \left(\frac{\iint (g+1)^{-1} Fh_i^2 \pi k_i^2 dgd\eta}{\iint Fh_i^2 \pi dgd\eta} - \frac{\iint (g+1)^{-1} Fh_i^2 \pi k_i^2 dgd\eta}{\iint Fh_i^2 \pi k_i^2 dgd\eta} \right)^2.$$

Hence we have

$$\frac{A(w, s; i)}{3} \leq A_1(w, s; i) + A_2(w, s; i) + A_3(w, s; i),$$

where

$$A_1(w, s; i) = \left\{ \iint \left| \frac{1}{\iint F\pi dgd\eta} - \frac{h_i^2}{\iint Fh_i^2 \pi dgd\eta} \right| \frac{F\pi dgd\eta}{g+1} \right\}^2 \iint Fh_i^2 \pi dgd\eta,$$

$$A_2(w, s; i) = \frac{\left(\iint (g+1)^{-1} Fh_i^2 \pi (1 - k_i^2) dgd\eta \right)^2}{\iint Fh_i^2 \pi dgd\eta},$$

$$A_3(w, s; i) = \frac{\left(\iint (g+1)^{-1} Fh_i^2 \pi k_i^2 dgd\eta \right)^2}{(\iint Fh_i^2 \pi dgd\eta)^2 \iint Fh_i^2 \pi k_i^2 dgd\eta} \left(\iint Fh_i^2 \pi (1 - k_i^2) dgd\eta \right)^2.$$

In Sects. A.7.1–A.7.3, we prove that

$$\lim_{i \to \infty} \iint \|x\|^2 s^{n/2-1} A_\ell(\|x\|^2/s, s; i) dx ds = 0, \quad \text{for } \ell = 1, 2, 3,$$

which implies that $\Delta_i \to 0$ as $i \to \infty$. Thus the corresponding generalized Bayes estimator is admissible among all estimators, as was to be shown.

3.5 Simple Bayes Estimators

Interestingly, and perhaps somewhat surprisingly, suitable choices of the constants a and b (with $c = 0$) lead to admissible minimax generalized Bayes estimators of a simple form. Further, this form represents a relatively minor adjustment to the form of the James–Stein estimator. Here are the details. Consider the case $b = n/2 - a$ in (3.38). For the numerator of (3.38), we have

$$\int_0^\infty \frac{(g+1)^{-p/2-a-1}\{g/(g+1)\}^b dg}{\{1 + w/(g+1)\}^{p/2+n/2+1}} = \int_0^\infty \frac{g^{n/2-a} dg}{(g+1+w)^{p/2+n/2+1}}$$

$$= \frac{1}{(1+w)^{p/2+a}} \int_0^\infty \frac{t^{n/2-a} dg}{(1+t)^{p/2+n/2+1}} = \frac{B(n/2+1-a, p/2+a)}{(1+w)^{p/2+a+2}}.$$

Similarly, for the denominator of of (3.38), we have

$$\int_0^\infty \frac{(g+1)^{-p/2-a}\{g/(g+1)\}^b dg}{\{1 + w/(g+1)\}^{p/2+n/2+1}}$$

$$= \int_0^\infty (1+g) \frac{(g+1)^{-p/2-1-a}\{g/(g+1)\}^b dg}{\{1 + w/(g+1)\}^{p/2+n/2+1}}$$

$$= \frac{B(n/2+1-a, p/2+a)}{(1+w)^{p/2+a}} + \frac{B(n/2+2-a, p/2-1+a)}{(1+w)^{p/2-1+a}}$$

$$= \frac{B(n/2+1-a, p/2+a)}{(1+w)^{p/2+a}} \left(1 + \frac{n/2+1-a}{p/2-1+a}(w+1)\right).$$

Thus the generalized Bayes estimator is of the form

$$\hat{\theta}_\alpha^{SB} = \left(1 - \frac{\alpha}{\|x\|^2/s + \alpha + 1}\right)x,$$

where $\alpha = (p/2-1+a)/(n/2+1-a)$. This estimator was discovered and studied in Maruyama and Strawderman (2005). By Theorem 3.6, provided

$$\alpha > \frac{p-2}{n+2} \quad \Leftrightarrow \quad a > 0,$$

$\hat{\theta}_\alpha^{SB}$ is admissible among all estimators. Also, by Theorem 3.5, $\hat{\theta}_\alpha^{SB}$ with $\alpha = (p - 2)/(n + 2)$ is admissible within the class of equivariant estimators. Additionally, by

Corollary 3.5, in Sect. 3.7 below, minimaxity of $\hat{\theta}_\alpha^{SB}$ holds for

$$0 < \alpha \leq 2\frac{p-2}{n+2} \quad \Leftrightarrow \quad -p/2+1 < a \leq \frac{(p-2)(n+2)}{2(2p+n-2)}.$$

3.6 Inadmissibility

3.6.1 A General Sufficient Condition for Inadmissibility

This section is devoted to the question of inadmissibility of shrinkage estimators of the form $\hat{\theta}_\phi = (1 - \phi(w)/w)x$ where $w = \|x\|^2/s$. Note that such estimators are equivariant. By (1.48) in Chap. 1, with $\psi(w) = \phi(w)/w$, the SURE for an estimator of the form $\hat{\theta}_\phi$ is

$$\hat{R}_\phi = p + \frac{(n+2)\{\phi(w) - 2c_{p,n}\}\phi(w)}{w} - 4\phi'(w)\{1 + \phi(w)\}, \qquad (3.49)$$

where $c_{p,n} = (p-2)/(n+2)$. For a competing estimator of the form

$$\hat{\theta}_{\phi+v} = \left(1 - \frac{\phi(w) + v(w)}{w}\right)x,$$

the difference in the SURE between $\hat{\theta}_\phi$ and $\hat{\theta}_{\phi+v}$ is

$$\hat{R}_\phi - \hat{R}_{\phi+v} = v(w)\{\Delta_1(w; \phi) + \Delta_2(w; \phi, v)\} \qquad (3.50)$$

where

$$\Delta_1(w; \phi) = 2(n+2)\frac{c_{p,n} - \phi(w)}{w} + 4\phi'(w),$$

$$\Delta_2(w; \phi, v) = -(n+2)\frac{v(w)}{w} + 4v'(w) + 4\frac{v'(w)}{v(w)}\{1 + \phi(w)\}.$$

Our approach to finding an estimator dominating $\hat{\theta}_\phi$ is to find a non-zero solution $v(w)$ to the differential inequality $\hat{R}_\phi - \hat{R}_{\phi+v} \geq 0$. Here is the result.

Theorem 3.7 (Maruyama and Strawderman 2017) *Let* $c_{p,n} = (p-2)/(n+2)$. *Suppose*

$$\limsup_{w\to\infty} \phi(w) \leq c_{p,n}$$
$$(3.51)$$
and $\liminf_{w\to\infty} \log w \{(n+2)\{c_{p,n} - \phi(w)\} + 2w\phi'(w)\} > 2(1 + c_{p,n}).$

Then the estimator $\hat{\theta}_\phi = (1 - \phi(w)/w)x$ *with* $w = \|x\|^2/s$ *is inadmissible.*

Proof By (3.51), there exist

$$w_1 > \exp(1) \text{ and } 0 < \epsilon < 1 \tag{3.52}$$

such that for all $w \geq w_1$,

$$\phi(w) - c_{p,n} \leq \frac{1 + c_{p,n}}{6} \epsilon$$

and

$$\log w \left\{ (n+2)\{c_{p,n} - \phi(w)\} + 2w\phi'(w) \right\} - 2(1 + c_{p,n})(1 + \epsilon) \geq 0,$$
$$\text{or equivalently} \quad \Delta_1(w; \phi) - 4 \frac{(1 + c_{p,n})(1 + \epsilon)}{w \log w} \geq 0, \tag{3.53}$$

Let $q(w; w_2)$ be the cumulative distribution function of $Y + w_2$, where $w_2 > w_1$ will be precisely determined later and Y is a Gamma random variable with the probability density function $y \exp(-y) I_{(0,\infty)}(y)$, that is,

$$q(w; w_2) = \begin{cases} 0 & \text{for } 0 \leq w < w_2 \\ \int_0^{w-w_2} y \exp(-y) dy & \text{for } w \geq w_2. \end{cases}$$

Then $q(w; w_2)$ is non-decreasing, differentiable with $q'(w)|_{w=w_2} = 0$ and $q(\infty) = 1$.

Let $v(w)$ for the competing estimator be given by

$$v(w; w_2) = \frac{q(w; w_2)}{(\log w)^{1+\epsilon/2}}, \tag{3.54}$$

with ϵ satisfying (3.52) and (3.53). Then, for all $w \geq w_2$, we have

$$\Delta_2[w; \phi, v(w; w_2)] + 4 \frac{(1 + c_{p,n})(1 + \epsilon)}{w \log w}$$

$$= -(n+2) \frac{q(w; w_2)}{w(\log w)^{1+\epsilon/2}} - \frac{4(1 + \epsilon/2) q(w; w_2)}{w(\log w)^{2+\epsilon/2}} + \frac{4q'(w; w_2)}{(\log w)^{1+\epsilon/2}}$$

$$+ 4 \left\{ \frac{q'(w; w_2)}{q(w; w_2)} - \frac{1 + \epsilon/2}{w \log w} \right\} \{1 + \phi(w)\} + 4 \frac{(1 + c_{p,n})(1 + \epsilon)}{w \log w}.$$

Note that $q'(w; w_2) \geq 0$, $q(w; w_2) \leq 1$, $(\log w)^{2+\epsilon/2} \geq (\log w)^{1+\epsilon/2}$ and

$$4(1 + \epsilon/2)\{1 + \phi(w)\} \leq 4(1 + \epsilon/2)\left(1 + c_{p,n} + \frac{1 + c_{p,n}}{6} \epsilon\right)$$

$$= (1 + c_{p,n})\left(4 + 2\epsilon + \frac{2}{3}(1 + \epsilon/2)\epsilon\right) \leq (4 + 3\epsilon)(1 + c_{p,n}).$$

Hence

$$\Delta_2[w; \phi, \nu(w; w_2)] + 4\frac{(1 + c_{p,n})(1 + \epsilon)}{w \log w}$$

$$\geq -\frac{4(1 + c_{p,n})(1 + 3\epsilon/4)}{w \log w} - \frac{4(1 + \epsilon/2) + n + 2}{w(\log w)^{1+\epsilon/2}} + 4\frac{(1 + c_{p,n})(1 + \epsilon)}{w \log w}$$

$$= \frac{(1 + c_{p,n})\epsilon}{w \log w}\left(1 - \frac{4(1 + \epsilon/2) + n + 2}{(1 + c_{p,n})\epsilon}\frac{1}{(\log w)^{\epsilon/2}}\right)$$

$$\geq \frac{(1 + c_{p,n})\epsilon}{w \log w}\left(1 - \frac{4(1 + \epsilon/2) + n + 2}{\epsilon}\frac{1}{(\log w)^{\epsilon/2}}\right). \tag{3.55}$$

Now let

$$w_2 = \max\left\{\exp\left(\left\{\frac{4(1 + \epsilon/2) + n + 2}{\epsilon}\right\}^{2/\epsilon}\right), w_1\right\}.$$

Then, by (3.53) and (3.55), we have

$$\Delta_1(w; \phi) + \Delta_2[w; \phi, \nu(w; w_2)]$$

$$= \left\{\Delta_1(w; \phi) - \frac{(1 + c_{p,n})(1 + \epsilon)}{(1/4)w \log w}\right\} + \left\{\Delta_2[w; \phi, \nu(w; w_2)] + \frac{(1 + c_{p,n})(1 + \epsilon)}{(1/4)w \log w}\right\}$$

$$\geq 0, \tag{3.56}$$

for all $w \geq w_2$. Hence, by (3.50), (3.54) and (3.56),

$$\hat{R}_\phi - \hat{R}_{\phi+\nu} = \nu(w)\{\Delta_1(w; \phi) + \Delta_2(w; \phi, \nu(w; w_2))\}\begin{cases} = 0 & \text{for } w < w_2 \\ \geq 0 & \text{for } w \geq w_2, \end{cases}$$

which completes the proof. $\qquad\square$

As a corollary of Theorem 3.7, we have the following result.

Corollary 3.2 *The estimator $\hat{\theta}_\phi$ is inadmissible if $\phi(w)$ satisfies either*

$$\limsup_{w \to \infty} \phi(w) < \frac{p - 2}{n + 2} \quad \text{and} \quad \lim_{w \to \infty} w\phi'(w) = 0 \tag{3.57}$$

or

$$\lim_{w \to \infty} \phi(w) = \frac{p - 2}{n + 2}, \quad \lim_{w \to \infty} w \log w \frac{\phi'(w)}{\phi(w)} = 0,$$

$$\text{and} \quad \liminf_{w \to \infty} \log w\left\{\frac{p - 2}{n + 2} - \phi(w)\right\} > \frac{2(p + n)}{(n + 2)^2}. \tag{3.58}$$

3.6.2 Inadmissible Generalized Bayes Estimators

In this subsection, we apply the results of the previous subsection to a class of generalized Bayes estimators. As in Sect. 2.5, we assume that $\Pi(dg)$ in (3.27) has a regularly varying density $\pi(g) = (g + 1)^{-a}\xi(g)$ where $\xi(g)$ satisfies AS.1 and AS.2 given in the end of Sect. 2.1. The corresponding generalized Bayes estimator is of the form $(1 - \phi(w)/w)x$ where

$$\phi(w) = w\frac{\int_0^\infty (g + 1)^{-p/2-1-a}\{1 + w/(g + 1)\}^{-(p/2+n/2+1)}\xi(g)dg}{\int_0^\infty (g + 1)^{-p/2-a}\{1 + w/(g + 1)\}^{-(p/2+n/2+1)}\xi(g)dg},$$

In addition to AS.1 and AS.2, we assume the following mild assumptions on the asymptotic behaviors on $\xi(g)$;

A.S.5 $\limsup\limits_{g\to\infty}\left\{(g + 1)\log(g + 1)\dfrac{\xi'(g)}{\xi(g)}\right\}$ is bounded,

A.S.6 $\xi(g)$ is ultimately monotone i.e., $\xi(g)$ is monotone on (g_0, ∞) for some $g_0 > 0$.

Under AS.1, AS.2, AS.5 and AS.6, we have the following result on the properties of $\phi(w)$.

Lemma 3.2 *Suppose* $-p/2 + 1 < a < n/2 + 1$. *Assume AS.1, AS.2, AS.5 and AS.6. Then $\phi(w)$ satisfies the following;*

1. $\lim\limits_{w\to\infty} \dfrac{\int_0^\infty (g + 1)^{-p/2-a}\{1 + w/(g + 1)\}^{-(p/2+n/2+1)}\xi(g)dg}{w^{-p/2+1-a}\xi(w)\mathrm{B}(p/2 - 1 + a, n/2 - a + 2)} = 1.$

2. $\lim\limits_{w\to\infty} \phi(w) = \dfrac{p/2 - 1 + a}{n/2 + 1 - a}.$

3. $\lim\limits_{w\to\infty} w\dfrac{\phi'(w)}{\phi(w)} = 0.$

Proof See Sect. A.10. □

By (3.57) of Corollary 3.2 and Parts 2 and 3 of Lemma 3.2, we have the following result.

Theorem 3.8 *Assume AS.1, AS.2, AS.5 and AS.6. Then the generalized Bayes estimator, with respect to the regularly varying density $\pi(g) = (g + 1)^{-a}\xi(g)$, is inadmissible if $-p/2 + 1 < a < 0$.*

As in Sect. 2.4.3, suppose $\Pi(dg)$ in (3.27) has a regularly varying density $\pi(g; a, b, c)$ as given in (3.35). It is easily seen that $\xi(g) = \{g/(g + 1)\}^b\{\log(g + 1) + 1\}^{-c}$, for $b > -1$ and $c \in \mathbb{R}$, satisfies AS.5 and AS.6 as well as AS.1, AS.2. Hence we have the following corollary.

Corollary 3.3 *Assume*

$$-p/2 + 1 < a < 0,\ b > -1,\ c \in \mathbb{R},$$

in $\pi(g; a, b, c)$. Then the corresponding generalized Bayes estimator is inadmissible.

When $\lim_{w \to \infty} \phi(w) = (p-2)/(n+2)$, recall that a sufficient condition for inadmissibility is given by (3.58) of Corollary 3.2. The following lemma on the behavior of $\phi(w)$ is helpful for providing an inadmissibility result for $\pi(g; a, b, c)$ when $a = 0$.

Lemma 3.3 *Let $a = 0$, $b > -1$, and $c \neq 0$ in $\pi(g; a, b, c)$. Then*

$$\lim_{w \to \infty} \log w \left\{ \frac{p-2}{n+2} - \phi(w) \right\} = -c \frac{2(p+n)}{(n+2)^2}, \tag{3.59}$$

$$and \quad -\lim_{w \to \infty} w \log w \frac{\phi'(w)}{\phi(w)} = 0. \tag{3.60}$$

Proof See Sect. A.11. □

Then, by Parts 2 and 3 of Lemma 3.2, Lemma 3.3, and (3.58) of Corollary 3.2, we have the following result.

Corollary 3.4 *Assume*

$$a = 0, \ b > -1, \ c < -1$$

in $\pi(g; a, b, c)$. Then the corresponding generalized Bayes estimator is inadmissible.

Note that Corollaries 3.3 and 3.4 correspond to Corollary 2.1 for the known scale case.

3.7 Minimaxity

3.7.1 A Sufficient Condition for Minimaxity

In this section, we study the minimaxity of shrinkage estimators of the form

$$\hat{\theta}_\phi = \left(1 - \frac{\phi(w)}{w} \right) x$$

where $w = \|x\|^2/s$ and $\phi(w)$ is differentiable. The risk function of the estimator is

$$R(\hat{\theta}_\phi; \theta, \eta) = p - 2 \sum_{i=1}^{n} E\left[\eta \frac{\phi(W)}{W} X_i (X_i - \theta_i) \right] + \eta E\left[S \frac{\phi^2(W)}{W} \right]. \tag{3.61}$$

As in (3.49) the SURE for an estimator $\hat{\theta}_\phi$ is give by $R(\hat{\theta}_\phi; \theta, \eta) = E[\hat{R}_\phi(W)]$, where

$$\hat{R}_\phi(w) = p + \frac{\{(n+2)\phi(w) - 2(p-2)\}\phi(w)}{w} - 4\phi'(w)\{1 + \phi(w)\}. \tag{3.62}$$

Hence, for a nonnegative $\phi(w)$, we have the following equivalence,

$$\frac{w\{\hat{R}_\phi(w) - p\}}{\phi(w)\{1 + \phi(w)\}} \leq 0 \quad \Leftrightarrow \quad \frac{2(p-2) - (n+2)\phi(w)}{1 + \phi(w)} + 4w\frac{\phi'(w)}{\phi(w)} \geq 0.$$

This implies the following result, which is Lemma 4.1 of Wells and Zhou (2008).

Theorem 3.9 *Assume that for $p \geq 3$ and a constant $\gamma \geq 0$, the differentiable function $\phi(w)$ satisfies the conditions: for any $w \geq 0$*

$$\frac{w\phi'(w)}{\phi(w)} \geq -\gamma \quad and \quad 0 \leq \phi(w) \leq 2\frac{p-2-2\gamma}{n+2+4\gamma}.$$

Then, the estimator $\hat{\theta}_\phi$ is minimax.

Kubokawa (2009) proposed an alternative expression for the risk function which differs from the SURE estimator given by (3.62). We will use the result below to strengthen Theorem 3.9.

Theorem 3.10 (Kubokawa 2009) *The risk function is $R(\hat{\theta}_\phi; \theta, \eta) = p + \eta \, E\,[(S/W)\mathcal{I}(W)]$, where*

$$\mathcal{I}(w) = \phi^2(w) + 2\phi(w) - (n+p)\int_0^1 z^{n/2}\phi(w/z)\mathrm{d}z.$$

Proof Unlike the development of (3.62), we apply both Lemmas 1.1 and 1.2 to the second term on the right hand side of (3.61). Define a function $\Phi(W)$ by

$$\Phi(w) = \frac{1}{2w}\int_0^1 z^{n/2}\phi(w/z)\mathrm{d}z = \frac{w^{n/2}}{2}\int_w^\infty \frac{\phi(t)}{t^{n/2+2}}\mathrm{d}t,$$

where the third expression results from the transformation $t = w/z$. Using Lemma 1.2, we obtain

$$\eta \, E^{S|X}\,[\Phi(W)S] = E^{S|X}\left[n\Phi(W) + 2S\frac{\partial}{\partial s}\Phi(W)\right] = E^{S|X}\left[\frac{\phi(W)}{W}\right], \quad (3.63)$$

where $E^{S|X}[\cdot]$ denotes the conditional expectation with respect to S given X. Note that all the expectations are finite since $\phi(w)$ is bounded.

By (3.63), we can rewrite the cross product term in (3.61) as

$$\eta\sum_{i=1}^p E\left[\frac{\phi(W)}{W}X_i(X_i - \theta_i)\right] = \eta^2\sum_{i=1}^p E\,[S\Phi(W)X_i(X_i - \theta_i)]. \quad (3.64)$$

Note

$$\frac{\partial}{\partial x_i}x_i\Phi(\|x\|^2/s) = \Phi(\|x\|^2/s) + 2\frac{x_i^2}{s}\Phi'(w)\bigg|_{w=\|x\|^2/s}, \quad (3.65)$$

where

$$\Phi'(w) = \frac{1}{2}\left(\frac{n}{2}w^{n/2-1}\int_w^\infty \frac{\phi(t)}{t^{n/2+2}}dt - \frac{\phi(w)}{w^2}\right) = \frac{1}{2}\left(n\frac{\Phi(w)}{w} - \frac{\phi(w)}{w^2}\right). \quad (3.66)$$

By Lemma 1.1, (3.65) and (3.66), we have

$$\eta \sum_{i=1}^p \mathrm{E}^{X|S}[\Phi(W)X_i(X_i - \theta_i)] = \mathrm{E}^{X|S}\left[(p+n)\Phi(W) - \frac{\phi(W)}{W}\right]$$

and, by (3.64)

$$\eta \sum_{i=1}^p \mathrm{E}\left[\frac{\phi(W)}{W}X_i(X_i - \theta_i)\right] = \eta\, \mathrm{E}\left[S\left\{(p+n)\Phi(W) - \frac{\phi(W)}{W}\right\}\right].$$

The proof is completed by combining the appropriate terms above. □

Suppose $\phi(w)$ is differentiable in Theorem 3.10. Then we have

$$\phi(w/z) - z^\gamma \phi(w) = \frac{z^\gamma}{w^\gamma}\{(w/z)^\gamma \phi(w/z) - w^\gamma \phi(w)\}$$

$$= \frac{z^\gamma}{w^\gamma}\int_w^{w/z}\left\{\frac{\mathrm{d}}{\mathrm{d}t}t^\gamma \phi(t)\right\}dt = \frac{z^\gamma}{w^\gamma}\int_w^{w/z} t^{\gamma-1}\phi(t)\left\{\gamma + t\frac{\phi'(t)}{\phi(t)}\right\}dt,$$

and hence

$$\mathcal{I}(w) = \phi^2(w) + 2\phi(w) - (n+p)\int_0^1 z^{n/2}\{\phi(w/z) - z^\gamma \phi(w) + z^\gamma \phi(w)\}\,dz$$

$$\leq \phi^2(w) + 2\phi(w) - (n+p)\phi(w)\int_0^1 z^{n/2+\gamma}dz$$

$$= \phi^2(w) + 2\phi(w) - 2\frac{n+p}{n+2+2\gamma}\phi(w)$$

$$= \phi(w)\left(\phi(w) - 2\frac{p-2-2\gamma}{n+2+2\gamma}\right),$$

where the inequality follows if $\phi(w) \geq 0$ and $w\phi'(w)/\phi(w) + \gamma \geq 0$. Then we have the following result.

Theorem 3.11 (Kubokawa 2009) *Assume that for $p \geq 3$ and a constant $\gamma \geq 0$, the differentiable function $\phi(w)$ satisfies the conditions for any $w \geq 0$,*

$$\frac{w\phi'(w)}{\phi(w)} \geq -\gamma \quad and \quad 0 \leq \phi(w) \leq 2\frac{p-2-2\gamma}{n+2+2\gamma}.$$

Then, the estimator $\hat{\theta}_\phi$ is minimax.

Note that the result given by Theorem 3.11 is slightly stronger than that in Theorem 3.9 since

$$2\frac{p-2-2\gamma}{n+2+4\gamma} \le 2\frac{p-2-2\gamma}{n+2+2\gamma}.$$

For this reason we will use Theorem 3.11, to consider the minimaxity of generalized Bayes estimator in Sect. 3.7.2.

3.7.2 Minimaxity of Some Generalized Bayes Estimators

Suppose $\pi(g) = (g+1)^{-a}\xi(g)$ where $\xi(g)$ satisfies AS.1–AS.4 as in Sect. 2.5.1. In this section, we investigate minimaxity of the corresponding generalized Bayes estimators with

$$\phi(w) = w\frac{\int_0^\infty (g+1)^{-p/2-a-1}\{1+w/(g+1)\}^{-(p/2+n/2+1)}\xi(g)dg}{\int_0^\infty (g+1)^{-p/2-a}\{1+w/(g+1)\}^{-(p/2+n/2+1)}\xi(g)dg}.$$

Recall that, in Sect. 2.5.1, $\Xi(g)$, $\Xi_1(g)$, $\Xi_2(g)$ and Ξ_{2*} were defined based on $\xi(g)$ and that the properties of these functions are summarized in Lemma 2.1. These results imply to the following properties for $\phi(w)$.

Lemma 3.4 *Suppose* $-p/2+1 < a < n/2+1 - \Xi_{2*}$. *Then*

$$\phi(w) \le \frac{p-2+2a+2\Xi_{2*}}{n+2-2a-2\Xi_{2*}} \quad and \quad w\frac{\phi'(w)}{\phi(w)} \ge -\Xi_{2*}. \tag{3.67}$$

Proof Section A.12.   □

Hence by Theorem 3.11 and Lemma 3.4, we have the following result.

Theorem 3.12 *The generalized Bayes estimator is minimax if*

$$\frac{p+2+2a+2\Xi_{2*}}{n-2-2a-2\Xi_{2*}} \le 2\frac{p-2-2\Xi_{2*}}{n+2+2\Xi_{2*}}.$$

For $\xi(g) = \{g/(g+1)\}^b/\{\log(g+1)+1\}^c$ with $b \ge 0$, the following corollary follows from Lemma 2.2 and Theorem 3.12.

Corollary 3.5 *For* $\pi(g; a, b, c)$ *given by (3.35) with* $b \ge 0$, *the corresponding generalized Bayes estimator is minimax if either*

$$-p/2+1 < a \le \frac{(p-2)(n+2)}{2(2p+n-2)}, \quad c \le 0$$

or

$$- p/2 + 1 < a < + \frac{(p-2)(n+2)}{2(2p+n-2)}, \quad c > 0,$$

$$\frac{(p-2+2a)\{1 + \log(b/c + 1)\} + 2c}{(n+2-2a)\{1 + \log(b/c + 1)\} - 2c} \leq 2 \frac{(p-2)\{1 + \log(b/c + 1)\} - 2c}{(n+2)\{1 + \log(b/c + 1)\} + 2c}.$$

Suppose

$$\xi(g) = \left(\frac{g}{g+1}\right)^b \quad \text{for } -1 < b < 0,$$

as considered in Sect. 2.5.2. For this case, the behavior of the corresponding $\phi(w)$ is summarized in the next result.

Lemma 3.5 *Let* $-1 < b < 0$. *Then* $\phi(w)$ *of the corresponding generalized Bayes estimator satisfies*

$$\phi(w) \leq \frac{p - 2 + 2a}{n + 2 - 2a + b(p+n)} \quad \text{and} \quad w\frac{\phi'(w)}{\phi(w)} \geq \frac{(p+2a)b}{2(b+1)}.$$

Proof Section A.13.   □

Thus Theorem 3.11 and Lemma 3.5, give minimaxity under the following conditions.

Theorem 3.13 *The generalized Bayes estimator is minimax if* $-1 < b < 0$ *and*

$$\frac{p - 2 + 2a}{n + 2 - 2a + b(p+n)} \leq 2\frac{(p-2)(b+1) + b(p+2a)}{(n+2)(b+1) - b(p+2a)}.$$

3.8 Improvement on the James–Stein Estimator

In this section we extend the discussion in Sect. 2.6 to the case of unknown variance. As in (1.49) and Theorem 1.8, the James–Stein estimator

$$\hat{\theta}_{JS} = \left(1 - \frac{p-2}{n+2}\frac{S}{\|X\|^2}\right)X$$

dominates the estimator X for $p \geq 3$. Using the expression for the risk of $\hat{\theta}_\phi$ given by (3.49), the risk difference is given by

$$\Delta(\lambda) = R(\hat{\theta}_{JS}; \theta, \eta) - R(\hat{\theta}_\phi; \theta, \eta)$$

$$= E\left[-(n+2)\frac{\{\phi(W) - c_{p,n}\}^2}{W} + 4\{1 + \phi(W)\}\phi'(W)\right],$$

where $c_{p,n} = (p-2)/(n+2)$, $\lambda = \eta\|\theta\|^2$ and $W = \|X\|^2/S$. Conditions on ϕ which ensure that $\Delta(\lambda) \geq 0$ are provided in the following theorem.

Theorem 3.14 (Kubokawa 1994) *The shrinkage estimator $\hat{\theta}_\phi$ improves on the James–Stein estimator $\hat{\theta}_{JS}$ if ϕ satisfies the following conditions: (i) $\phi(w)$ is non-decreasing in w; (ii) $\lim_{w\to\infty}\phi(w) = (p-2)/(n+2)$ and $\phi(w) \geq \phi_0(w)$ where*

$$\phi_0(w) = w\frac{\int_0^\infty (g+1)^{-p/2-1}\{1+w/(g+1)\}^{-p/2-n/2-1}dg}{\int_0^\infty (g+1)^{-p/2}\{1+w/(g+1)\}^{-p/2-n/2-1}dg}.$$

Proof Let $U = \eta\|X\|^2$ and $V = \eta S$, and let $f_p(u; \lambda)$ and $f_n(v)$ be density functions of $\chi_p^2(\lambda)$ and χ_n^2, respectively. Then $U \sim \chi_p^2(\lambda)$ where $\lambda = \eta\|\theta\|^2$ and $V \sim \chi_n^2$. The expected value of a function $\psi(\|x\|^2/s)$ may be expressed as

$$E[\psi(\|X\|^2/S)] = E[\psi(\{\eta\|X\|^2\}/\{\eta S\})]$$

$$= \iint \psi(u/v)f_p(u; \lambda)f_n(v)dudv = \iint \psi(w)vf_p(wv; \lambda)f_n(v)dvdw$$

$$= \int_0^\infty \psi(w)\left\{\int_0^\infty vf_p(wv; \lambda)f_n(v)dv\right\}dw$$

$$= \int_0^\infty \psi(w)\sum_{i=0}^\infty \frac{(\lambda/2)^i}{e^{\lambda/2}i!}\left\{\int_0^\infty v\frac{(wv)^{p/2+i-1}\exp(-wv/2)}{\Gamma(p/2+i)2^{p/2+i}}\frac{v^{n/2-1}\exp(-v/2)}{\Gamma(n/2)2^{n/2}}dv\right\}dw$$

$$= \int_0^\infty \psi(w)j_{p,n}(w; \lambda)dw,$$

where

$$j_{p,n}(w; \lambda) = \sum_{i=0}^\infty \frac{(\lambda/2)^i}{e^{\lambda/2}i!}\frac{w^{p/2-1}(1+w)^{-p/2-n/2}}{B(p/2+i, n/2)}\left(\frac{w}{w+1}\right)^i.$$

Then, arguing as in (2.45) and (2.46), the first term of $\Delta(\lambda)$ may be expressed as written as

$$-E\left[W^{-1}(\phi(W) - c_{p,n})^2\right]$$

$$= 2\int_0^\infty \{\phi(w) - c_{p,n}\}\phi'(w)\left\{\int_0^\infty \frac{j_{p,n}(w/(g+1); \lambda)}{g+1}dg\right\}dw$$

and hence $\Delta(\lambda)$ may be written as

$$\Delta(\lambda) = 2 \int_0^\infty \phi'(w) \Big((n+2)\{\phi(w) - c_{p,n}\} \int_0^\infty \frac{j_{p,n}(w/(g+1); \lambda)}{g+1} dg$$

$$+ 2\{1 + \phi(w)\} j_{p,n}(w; \lambda) \Big) dw$$

$$= 2 \int_0^\infty \phi'(w) \Big((n+2)\{\phi(w) - c_{p,n}\} + 2\{1 + \phi(w)\} J_{p,n}(w; \lambda) \Big)$$

$$\times \Big\{ \int_0^\infty \frac{j_{p,n}(w/(g+1); \lambda)}{g+1} dg \Big\} dw,$$

where

$$J_{p,n}(w; \lambda) = \frac{j_{p,n}(w; \lambda)}{\int_0^\infty (g+1)^{-1} j_{p,n}(w/(g+1); \lambda) dg}.$$

Further, as in (2.47), $J_{p,n}(w; \lambda) \geq J_{p,n}(w; 0)$ holds where

$$J_{p,n}(w; 0) = \frac{j_{p,n}(w; 0)}{\int_0^\infty (g+1)^{-1} j_{p,n}(w/(g+1); 0) dg}$$

$$= \frac{(1+w)^{-p/2-n-2}}{\int_0^\infty (g+1)^{-p/2} \{1 + w/(g+1)\}^{-p/2-n/2} dg}. \tag{3.68}$$

Hence we have $\Delta(\lambda) \geq 0$ if $\phi'(w) \geq 0$ and

$$(n+2)\{\phi(w) - c_{p,n}\} + 2\{1 + \phi(w)\} J_{p,n}(w; 0) \geq 0,$$

which is equivalent to $\phi(w) \geq \phi_0(w)$ where

$$\phi_0(w) = \frac{p - 2 - 2J_{p,n}(w; 0)}{n + 2 + 2J_{p,n}(w; 0)} \tag{3.69}$$

$$= \frac{(p-2) \int_0^\infty \frac{(g+1)^{-p/2} dg}{\{1 + w/(g+1)\}^{p/2+n/2}} - \frac{2}{(1+w)^{p/2+n/2}}}{(n+2) \int_0^\infty \frac{(g+1)^{-p/2} dg}{\{1 + w/(g+1)\}^{p/2+n/2}} + \frac{2}{(1+w)^{p/2+n/2}}}. \tag{3.70}$$

For the denominator of (3.70), an integration by parts gives

$$(n+2) \int_0^\infty \frac{(g+1)^{-p/2} dg}{\{1 + w/(g+1)\}^{p/2+n/2}} + \frac{2}{(1+w)^{p/2+n/2}}$$

$$= (n+2) \int_0^\infty \frac{(g+1)^{n/2} dg}{(1+g+w)^{p/2+n/2}} + \frac{2}{(1+w)^{p/2+n/2}}$$

$$= 2 \int_0^\infty (g+1)^{n/2+1} \left\{ \frac{(p+n)/2}{(1+g+w)^{p/2+n/2+1}} \right\} dg$$

$$= (p+n) \int_0^\infty \frac{(g+1)^{-p/2} dg}{\{1 + w/(g+1)\}^{p/2+n/2+1}}. \qquad (3.71)$$

Similarly, for the numerator of (3.70), an integration by parts gives

$$(p-2) \int_0^\infty \frac{(g+1)^{-p/2} dg}{\{1 + w/(g+1)\}^{p/2+n/2}} - \frac{2}{(1+w)^{p/2+n/2}}$$

$$= 2 \int_0^\infty (g+1)^{-p/2+1} \left\{ \frac{w}{(g+1)^2} \frac{(p+n)/2}{\{1 + w/(g+1)\}^{p/2+n/2+1}} \right\} dg$$

$$= (p+n) w \int_0^\infty \frac{(g+1)^{-p/2-1} dg}{\{1 + w/(g+1)\}^{p/2+n/2+1}}. \qquad (3.72)$$

By (3.70), (3.71) and (3.72), we have

$$\phi_0(w) = w \frac{\int_0^\infty (g+1)^{-p/2-1} \{1 + w/(g+1)\}^{-p/2-n/2-1} dg}{\int_0^\infty (g+1)^{-p/2} \{1 + w/(g+1)\}^{-p/2-n/2-1} dg}, \qquad (3.73)$$

which completes the proof of Theorem 3.14.   □

By (3.68), we have

$$J_{p,n}(w; 0) = \frac{1}{\int_0^\infty (g+1)^{n/2} \{(1+w)/(1+w+g)\}^{p/2+n/2} dg},$$

which is decreasing in w and approaches 0 as $w \to \infty$. It then follows directly from the first line of (3.69) that

$$\phi_0'(w) \geq 0, \quad \lim_{w \to \infty} \phi_0(w) = (p-2)/(n+2),$$

and hence the function $\phi_0(w)$ satisfies conditions (i) and (ii) of Theorem 3.14. It follows that the estimator associated with $\phi_0(w)$ is a minimax estimator improving on the James–Stein estimator. Further, comparing $\phi_0(w)$ with (3.36), we see that

$$\left(1 - \frac{\phi_0(\|X\|^2/S)}{\|X\|^2/S} \right) X$$

can be characterized as the generalized Bayes estimator under $\pi(g; a, b, c)$ in (3.35) with $a = b = c = 0$, or equivalently, the joint Stein (1974) prior given by (1.23),

$$\eta^{-1} \times \eta^{p/2} \pi_S(\eta \|\theta\|^2) = \eta^{-1} \times \eta^{p/2} \{\eta \|\theta\|^2\}^{1-p/2} = \|\theta\|^{2-p}, \qquad (3.74)$$

where π_S is given by (1.14).

Additionally, by (3.73), $\phi_0(w) \leq w$ and hence the the truncated function

$$\phi_{JS}^+ = \min\{w, (p-2)/(n+2)\}$$

corresponding to the James–Stein positive-part estimator

$$\hat{\theta}_{JS}^+ = \max\left(0, 1 - \frac{p-2}{n+2}\frac{S}{\|X\|^2}\right)X,$$

also satisfies conditions (i) and (ii) of Theorem 3.14, which implies that the James–Stein positive-part estimator dominates the James–Stein estimator, See Baranchik (1964) and Lehmann and Casella (1998) for the original proof of the domination.

It seems that the choice $a = b = c = 0$ in $\pi(g; a, b, c)$ is the only one which satisfies the conditions (i) and (ii) of Theorem 3.14. Recall, however, that we have concentrated on priors with $\nu = 1$ in (3.6) when deriving minimaxity and admissibility results in this chapter. As a choice of prior with $\nu \neq -1$ in (3.26), suppose the joint improper prior

$$\eta^{\alpha(n+p)/2-1} \times \int_0^\infty \frac{\eta^{p/2}}{(2\pi)^{p/2} g^{p/2}} \exp\left(-\frac{\eta\|\theta\|^2}{2g}\right) \frac{1}{(g+1)^{\alpha(p-2)/2}} dg,$$

for $\alpha > 0$. The choice $\alpha = 0$ corresponds to the joint Stein prior (3.74). Then the generalized Bayes estimator is given by

$$\hat{\theta}_\alpha = \left(1 - \frac{\int_0^\infty (g+1)^{-(\alpha+1)(p/2-1)-2}\{1 + w/(g+1)\}^{-(\alpha+1)(p/2+n/2)-1} dg}{\int_0^\infty (g+1)^{-(\alpha+1)(p/2-1)-1}\{1 + w/(g+1)\}^{-(\alpha+1)(p/2+n/2)-1} dg}\right)x.$$

The following result is due to Maruyama (1999).

Theorem 3.15 (Maruyama 1999) *The generalized Bayes estimator $\hat{\theta}_\alpha$ for $\alpha > 0$ dominates the James–Stein estimator $\hat{\theta}_{JS}$. Further $\hat{\theta}_\alpha$ approaches the James–Stein positive-part estimator $\hat{\theta}_{JS}^+$ as $\alpha \to \infty$.*

Proof Appendix A.14. □

We do not know whether $\hat{\theta}_\alpha$, for $\alpha > 0$, is admissible within the class of equivariant estimators.

References

Baranchik AJ (1964) Multiple regression and estimation of the mean of a multivariate normal distribution. Technical Report 51, Department of Statistics, Stanford University

Blyth CR (1951) On minimax statistical decision procedures and their admissibility. Ann Math Stat 22:22–42

Kubokawa T (1994) A unified approach to improving equivariant estimators. Ann Stat 22(1):290–299

Kubokawa T (2009) Integral inequality for minimaxity in the Stein problem. J Jpn Stat Soc 39(2):155–175

Lehmann EL, Casella G (1998) Theory of point estimation, 2nd edn. Springer Texts in Statistics, Springer, New York

Maruyama Y (1999) Improving on the James–Stein estimator. Stat Decis 17(2):137–140

Maruyama Y, Strawderman WE (2005) A new class of generalized Bayes minimax ridge regression estimators. Ann Stat 33(4):1753–1770

Maruyama Y, Strawderman WE (2017) A sharp boundary for SURE-based admissibility for the normal means problem under unknown scale. J Multivariate Anal 162:134–151

Maruyama Y, Strawderman WE (2020) Admissible Bayes equivariant estimation of location vectors for spherically symmetric distributions with unknown scale. Ann Stat 48(2):1052–1071

Maruyama Y, Strawderman WE (2021) Admissible estimators of a multivariate normal mean vector when the scale is unknown. Biometrika 108(4):997–1003

Maruyama Y, Strawderman WE (2023) On admissible estimation of a mean vector when the scale is unknown. Bernoulli 29(1):153–180

Stein C (1974) Estimation of the mean of a multivariate normal distribution. In: Proceedings of the prague symposium on asymptotic statistics (Charles University, Prague, 1973). vol. II. Charles University, Prague, pp 345–381

Strawderman WE (1973) Proper Bayes minimax estimators of the multivariate normal mean vector for the case of common unknown variances. Ann Stat 1:1189–1194

Wells MT, Zhou G (2008) Generalized Bayes minimax estimators of the mean of multivariate normal distribution with unknown variance. J Multivariate Anal 99(10):2208–2220

Appendix A
Miscellaneous Lemmas and Technical Proofs

A.1 Identities and Inequalities

In this section, we summarize some useful lemmas used in this book.

A.1.1 Identities

This identity, proved by by completing the square, is frequently used.

Lemma A.1 $\|x - \mu\|^2 + \dfrac{\|\mu\|^2}{i} = \dfrac{i+1}{i}\left\|\mu - \dfrac{i}{i+1}x\right\|^2 + \dfrac{\|x\|^2}{i+1}.$

Proof

$$\|x - \mu\|^2 + \frac{\|\mu\|^2}{i} = \frac{i+1}{i}\|\mu\|^2 - 2\mu^{\mathrm{T}}x + \|x\|^2$$

$$= \frac{i+1}{i}\left\|\mu - \frac{i}{i+1}x\right\|^2 - \frac{i}{i+1}\|x\|^2 + \|x\|^2 = \frac{i+1}{i}\left\|\mu - \frac{i}{i+1}x\right\|^2 + \frac{\|x\|^2}{i+1}.$$

$\square$

These standard results on multiple integrals are also often used in this book.

Lemma A.2 *1.* $\displaystyle\int_{\mathbb{R}^p} f(\|x\|^2)\mathrm{d}x = \frac{\pi^{p/2}}{\Gamma(p/2)}\int_0^\infty t^{p/2-1}f(t)\mathrm{d}t.$
 2. Assume $p/2 + \alpha > 0$ and $\beta > 0$. Then

$$\int_{\mathbb{R}^p} (\|x\|^2)^\alpha \exp\left(-\frac{\|x\|^2}{\beta}\right)\mathrm{d}x = \frac{\pi^{p/2}}{\Gamma(p/2)}\Gamma(p/2 + \alpha)\beta^{p/2+\alpha}.$$

© The Author(s), under exclusive license to Springer Nature Singapore Pte Ltd. 2023
Y. Maruyama et al., *Stein Estimation*, JSS Research Series in Statistics,
https://doi.org/10.1007/978-981-99-6077-4

3. *Assume $\beta > \alpha > -p/2$. Then*

$$\int_{\mathbb{R}^p} (\|x\|^2)^\alpha (1 + \|x\|^2)^{-p/2-\beta} dx = \frac{\pi^{p/2}}{\Gamma(p/2)} B(p/2 + \alpha, \beta - \alpha).$$

A.1.2 Inequalities

Different forms of the Cauchy–Schwarz inequality are of use in several places throughout the text.

Lemma A.3 (Cauchy–Schwarz inequality) *Assume $m(x) \geq 0$ on Ω.*

1. $\left\{ \int_\Omega f(x)g(x)m(dx) \right\}^2 \leq \int_\Omega \{f(x)\}^2 m(dx) \int_\Omega \{g(x)\}^2 m(dx).$
2. *Let $F(x) = (f_1(x), \ldots, f_p(x))^{\mathrm{T}}$. Then*

$$\left\| \int_\Omega F(x)g(x)m(dx) \right\|^2 \leq \int_\Omega \|F(x)\|^2 m(dx) \int_\Omega \{g(x)\}^2 m(dx).$$

3. $\left(\sum_{i=1}^p a_i \right)^2 \leq p \sum_{i=1}^p a_i^2.$

Proof [Part 1] For $t \in \mathbb{R}$, we have

$$\int_\Omega \{tf(x) + g(x)\}^2 m(dx)$$

$$= t^2 \int_\Omega \{f(x)\}^2 m(dx) + \int_\Omega \{g(x)\}^2 m(dx) + 2t \int_\Omega f(x)g(x)m(dx)$$

$$= \int_\Omega \{f(x)\}^2 m(dx) \left(t + \frac{\int_\Omega f(x)g(x)m(dx)}{\int_\Omega \{f(x)\}^2 m(dx)} \right)^2$$

$$+ \int_\Omega \{g(x)\}^2 m(dx) - \frac{\{\int_\Omega f(x)g(x)m(dx)\}^2}{\int_\Omega \{f(x)\}^2 m(dx)}.$$

Let $t_* = -\{\int_\Omega \{f(x)\}^2 m(dx)\}^{-1} \int_\Omega f(x)g(x)m(dx)$. Then

$$0 \leq \int_\Omega \{t_* f(x) + g(x)\}^2 m(dx) = \int_\Omega \{g(x)\}^2 m(dx) - \frac{\{\int_\Omega f(x)g(x)m(dx)\}^2}{\int_\Omega \{f(x)\}^2 m(dx)},$$

which completes the proof of Part 1.

[Part 2] By Part 1, we have

$$\left\{\int_\Omega f_i(x)g(x)m(dx)\right\}^2 \leq \int_\Omega \{f_i(x)\}^2 m(dx) \int_\Omega \{g(x)\}^2 m(dx)$$

for $i = 1, \ldots, p$ and hence

$$\sum_{i=1}^p \left\{\int_\Omega f_i(x)g(x)m(dx)\right\}^2 \leq \sum_{i=1}^p \int_\Omega \{f_i(x)\}^2 m(dx) \int_\Omega \{g(x)\}^2 m(dx)$$

$$\leq \int_\Omega \sum_{i=1}^p \{f_i(x)\}^2 m(dx) \int_\Omega \{g(x)\}^2 m(dx).$$

Then Part 2 follows.

[Part 3] Suppose $m(dx)$ is the counting measure at $x_1, x_2, \ldots, x_p$. Let

$$f(x_1) = a_1, \ f(x_2) = a_2, \ldots, \ f(x_p) = a_p, \ g(x_1) = g(x_2) = \cdots = g(x_p) = 1.$$

Then Part 1 gives Part 3. □

This standard correlation inequality is used frequently.

Lemma A.4 (Correlation inequality) *Suppose $f(x)$ and $g(x)$ are both monotone non-decreasing in x. Let X be a continuous random variable. Then*

$$E[f(X)g(X)] \geq E[f(X)]E[g(X)].$$

Proof Suppose $G = E[g(X)]$. Let x_* satisfy

$$g(x) \begin{cases} \leq G & x \leq x_* \\ \geq G & x > x_*. \end{cases}$$

Then

$$\begin{aligned} E[f(X)g(X)] &- E[f(X)]E[g(X)] = E[f(X)\{g(X) - G\}] \\ &= E[f(X)\{g(X) - G\}I_{(-\infty,x_*]}(X)] + E[f(X)\{g(X) - G\}I_{(x_*,\infty)}(X)] \\ &\geq E[f(x_*)\{g(X) - G\}I_{(-\infty,x_*]}(X)] + E[f(x_*)\{g(X) - G\}I_{(x_*,\infty)}(X)] \\ &= f(x_*)E[g(X) - G] = 0. \end{aligned}$$ □

These functional inequalities are also frequently useful.

Lemma A.5 *1. For $x \in (0, 1)$, $(1 - x)^\alpha \geq 1 - \max(1, \alpha)x$.*
2. For $x \geq 0$ and $\alpha \in (0, 1)$, $(x + 1)^\alpha \leq x^\alpha + 1$.
3. Let $x > 0$ and $\alpha > -1$. Then $\left(\dfrac{x}{x+1}\right)^\alpha \leq x^\alpha I_{(0,1]}(x) + 2I_{(1,\infty)}(x)$.

4. *For $x \in (0, 1)$ and $\alpha > 0$, $|\log x| \leq 1/(\alpha x^\alpha)$.*
5. *For any $\alpha > 0$ and $\beta > 0$ and all $x \in (0, \infty)$, $|\log x|^\beta \leq \dfrac{x^{\alpha\beta} + x^{-\alpha\beta}}{\alpha^\beta}$.*
6. *Let $f(x)$ for $x \in (0, \infty)$ be positive and differentiable and $\lim\inf_{x\to 0} f'(x)/f(x) > -\infty$. Then $f(0) < \infty$.*
7. *For any $\alpha > 0$ and $\beta > 0$ and all $x \in (0, \infty)$, $x^\alpha \exp(-\beta x) \leq (\alpha/\beta)^\alpha$.*

Proof [Part 1] For $\alpha \leq 1$, we have $(1 - x)^\alpha \geq 1 - x = 1 - \max(1, \alpha)x$, since $0 < 1 - x < 1$. For $\alpha > 1$, $(1 - x)^\alpha$ is convex in $x \in (0, 1)$ and the derivative of $(1 - x)^\alpha$ at $x = 0$ is $-\alpha$. Hence we have

$$(1 - x)^\alpha \geq 1 + (-\alpha)(x - 0) = 1 - \max(1, \alpha)x.$$

[Part 2] Let $f(x) = 1 + x^\alpha - (x + 1)^\alpha$. Then $f(0) = 0$ and

$$f'(x) = \alpha x^{\alpha-1} - \alpha(x + 1)^{\alpha-1} = \alpha x^{\alpha-1}\left\{1 - \left(\frac{x}{x + 1}\right)^{1-\alpha}\right\} \geq 0,$$

and the result follows.

[Part 3] For $\alpha \geq 0$, we have

$$\{x/(x + 1)\}^\alpha \leq x^\alpha I_{(0,1]}(x) + I_{(1,\infty)}(x)$$

and the result follows. For $\alpha \in (-1, 0)$,

$$\begin{aligned}
\left(\frac{x}{x + 1}\right)^\alpha &= \left(\frac{x + 1}{x}\right)^{-\alpha} I_{(0,1]}(x) + \left(\frac{x + 1}{x}\right)^{-\alpha} I_{(1,\infty)}(x) \\
&\leq \{1 + (1/x)^{-\alpha}\} I_{(0,1]}(x) + 2^{-\alpha} I_{(1,\infty)}(x) \\
&\leq x^\alpha I_{(0,1]}(x) + 2 I_{(1,\infty)}(x),
\end{aligned}$$

where the first inequality follows from Part 2. Thus Part 3 follows.

[Part 4] Recall $\log x^{-\alpha} \leq x^{-\alpha} - 1$ for $x > 0$. Then for $x \in (0, 1)$, we have

$$\alpha \log \frac{1}{x} \leq \frac{1}{x^\alpha} - 1 \text{ and } |\log x| \leq \frac{1}{\alpha x^\alpha},$$

for $\alpha > 0$. Then Part 4 follows.

[Part 5] For $x \in (1, \infty)$, we have $\log x^\alpha \leq x^\alpha - 1$,

$$\alpha \log x \leq x^\alpha - 1 \text{ and } |\log x| \leq x^\alpha/\alpha.$$

Then, together with Parts 4 and 5 follows.

[Part 6] By the assumption, there exists M such that $f'(x)/f(x) \geq M$ for all $0 \leq x \leq 1$. Thus we have

$$\int_x^1 \frac{f'(t)}{f(t)} dt \geq M \int_x^1 dt,$$

which implies that $f(x) \leq f(1)e^{M(x-1)}$. This completes the proof.
[Part 7]

$$x^\alpha \exp(-\beta x) = \exp\left(\alpha\left\{\log x - \frac{\beta}{\alpha}x\right\}\right) \leq \exp\left(\alpha\left\{\log\left(\frac{\beta}{\alpha}x\right) - \left(\frac{\beta}{\alpha}x\right) + \log\frac{\alpha}{\beta}\right\}\right)$$

$$\leq \exp\left(\alpha\left\{-1 + \log(\alpha/\beta)\right\}\right) \leq (\alpha/\beta)^\alpha. \qquad \square$$

A.2 Some Properties of k_i

In this section, we investigate the function

$$k_i(g) = 1 - \frac{\log(g+1)}{\log(g+1+i)}.$$

which is used for proving admissibility of X for $p = 1$ and 2 through the Blyth method in Sects. 1.5 and 1.6.

Lemma A.6 *1.* $k_i(g) \leq \dfrac{(1+i)\log(1+i)}{(g+1+i)\log(g+1+i)}$ *for fixed* $i \geq 1$.

2. $\displaystyle\int_0^\infty k_i^2(g)dg \leq 1+i$.

3. $k_i'(g) \leq 0$ *for all* $g \geq 0$ *and* $\{k_i'(g)\}^2 \leq \dfrac{1}{\{(g+1)\log(g+1+i)\}^2}$.

4. $\displaystyle\int_0^\infty (g+1)\{k_i'(g)\}^2 dg \leq \dfrac{2}{\log(1+i)}$.

5. Assume $a \geq 0$, $b > -1$ *and* $c = 0$ *in* $\pi(g; a, b, c)$ *given in (2.7). Then*
$$\int_0^\infty \pi(g; a, b, 0)k_i^2(g)dg \leq \frac{1}{b+1} + 2(1+i).$$

Proof [Part 1] The function $k_i(g)$ may be rewritten as

$$k_i(g) = \frac{i\zeta(i/(g+1+i))}{(g+1+i)\log(g+1+i)},$$

where $\zeta(x) = -x^{-1}\log(1-x) = 1 + \sum_{l=1}^\infty x^l/(l+1)$, which is increasing in x.

Hence

$$k_i(g) \leq \frac{i\zeta(i/(1+i))}{(g+1+i)\log(g+1+i)} = \frac{(1+i)\log(1+i)}{(g+1+i)\log(g+1+i)}.$$

[Part 2] By Part 1,

$$\int_0^\infty k_i^2(g)dg \leq \int_0^\infty \frac{(1+i)^2\{\log(1+i)\}^2 dg}{(g+1+i)^2\{\log(g+1+i)\}^2} \leq \int_0^\infty \frac{(1+i)^2 dg}{(g+1+i)^2} = 1+i.$$

Then the result follows.

[Part 3] The derivative is

$$k_i'(g) = -\frac{1}{(g+1)\log(g+1+i)} + \frac{\log(g+1)}{(g+1+i)\{\log(g+1+i)\}^2}.$$

Then we have

$$
\begin{aligned}
\log(g+1+i)k_i'(g) &= -\frac{1}{g+1} + \frac{\log(g+1)}{(g+1+i)\log(g+1+i)} \\
&= -\frac{1}{g+1} + \frac{1-k_i(g)}{g+1+i}, \quad\quad\quad\quad\quad\quad (A.1)
\end{aligned}
$$

where

$$\frac{1}{g+1} > \frac{1-k_i(g)}{g+1+i} > 0. \quad\quad\quad\quad\quad\quad (A.2)$$

Hence $k_i'(g) \leq 0$.

By (A.1) and (A.2), we have $h_i(g) \leq 0$ and

$$\{\log(g+1+i)k_i'(g)\}^2 \leq \frac{1}{(g+1)^2}.$$

[Part 4] By Part 3, we have

$$(g+1)\{k_i'(g)\}^2 \leq \frac{1}{(g+1)\{\log(g+1+i)\}^2}.$$

Then we have

$$\int_0^i (g+1)\{k_i'(g)\}^2 dg \leq \frac{1}{\{\log(1+i)\}^2} \int_0^i \frac{dg}{g+1} = \frac{1}{\log(1+i)}$$

and

$$\int_i^\infty (g+1)\{k_i'(g)\}^2 dg \le \int_i^\infty \frac{dg}{(g+1)\{\log(g+1)\}^2} = \frac{1}{\log(1+i)}.$$

Then the result follows.

[Part 5] By assumption, $\pi(g; a, b) \le \{g/(g+1)\}^b$. Further by Part 3 of Lemma A.5,

$$\pi(g) \le \left(\frac{g}{g+1}\right)^b \le g^b I_{(0,1)}(g) + 2I_{(1,\infty)}(g). \tag{A.3}$$

Note $k_i^2(g) \le 1$ by definition. Hence, by (A.3) and Part 1 of this lemma, we have

$$\int_0^\infty \pi(g)k_i^2(g)dg \le \int_0^1 g^b dg + 2\int_0^\infty k_i^2(g)dg \le \frac{1}{b+1} + 2(1+i),$$

which completes the proof of Part 5. □

A.3 Some Properties Under $\pi(g; a, b, c)$

This section gives properties of functions related to $\pi(g; a, b, c)$, given in (2.7), and used throughout the text.

Lemma A.7 $\sup_x \|\nabla_x \log m_\pi(\|x\|^2; a, b, c)\|^2 < \infty.$

Proof By (2.6), let

$$f(\|x\|^2) = \|\nabla_x \log m_\pi(\|x\|^2; a, b, c)\| \tag{A.4}$$

$$= \|x\| \frac{\int_0^\infty (g+1)^{-p/2-1} \exp\left(-\|x\|^2/\{2(g+1)\}\right) \pi(g; a, b, c)dg}{\int_0^\infty (g+1)^{-p/2} \exp\left(-\|x\|^2/\{2(g+1)\}\right) \pi(g; a, b, c)dg},$$

where

$$m_\pi(\|x\|^2; a, b, c) = \int_0^\infty \frac{(g+1)^{-p/2}}{(2\pi)^{p/2}} \exp\left(-\frac{\|x\|^2}{2(g+1)}\right) \pi(g; a, b, c)dg.$$

Clearly $f(0) = 0$ since

$$f(0) = 0 \times \frac{\int_0^\infty (g+1)^{-p/2-1} \pi(g; a, b, c)dg}{\int_0^\infty (g+1)^{-p/2} \pi(g; a, b, c)dg} = 0.$$

Note, as in (2.11),

$$\lim_{t \to \infty} \frac{t^{p/2-1} m_\pi(t; a, b, c)}{\pi(t; a, b, c)} = \frac{\Gamma(p/2 - 1 + a) 2^{p/2-1+a}}{(2\pi)^{p/2}}. \tag{A.5}$$

Similarly,

$$\lim_{t \to \infty} \frac{t^{p/2} \int_0^\infty \dfrac{\pi(g; a, b, c) dg}{(g+1)^{p/2+1} \exp(t/2(g+1))}}{\pi(t; a, b, c)} = \Gamma(p/2 + a) 2^{p/2+a}. \tag{A.6}$$

Hence, by (A.4), (A.5) and (A.6), we have $\lim_{t \to \infty} t^{1/2} f(t) = 2(p/2 - 1 + a)$, which implies that $\lim_{t \to \infty} f(t) = 0$. Together with $f(0) = 0$, $f(t)$ is bounded. $\square$

Lemma A.8 *1. For either $a < -$ or $\{a = 0$ and $c < -1\}$, $\int_1^\infty \dfrac{g^{-1} dg}{\pi(g; a, b, c)} < \infty$.*

2. For either $a > 0$ or $\{a = 0$ and $c > 1\}$, $\int_0^\infty \dfrac{\pi(g; a, b, c)}{g+1} dg < \infty$.

3. For either $a > 0$ or $\{a = 0$ and $c \geq -1\}$, $\int_1^\infty \dfrac{dg}{g\pi(g; a, b, c)} = \infty$.

Proof [Part 1] Let $\gamma = 2$ for $a < 0$ and $\gamma = -c$ for $a = 0$. Further let

$$f_1(g) = \frac{\{\log(g+1) + 1\}^{\gamma+c}}{(g+1)^{-a}} \left(\frac{g+1}{g}\right)^{b+1},$$

which is bounded for $g \in (1, \infty)$, for either $a < 0$ or $\{a = 0$ and $c < -1\}$, since $f_1(1) < \infty$ and $f_1(\infty) < \infty$. Then

$$\int_1^\infty \frac{dg}{g\pi(g; a, b, c)} = \int_1^\infty \frac{f_1(g) dg}{(g+1)\{\log(g+1) + 1\}^\gamma}$$

$$\leq \max_{g \geq 1} f_1(g) \int_1^\infty \frac{dg}{(g+1)\{\log(g+1) + 1\}^\gamma} = \max_{g \geq 1} f_1(g) \frac{(\log 2 + 1)^{1-\gamma}}{\gamma - 1},$$

which completes the proof of Part 1.
   [Part 2] For $0 < g < 1$,

$$\frac{\pi(g; a, b, c)}{g+1} = \frac{g^b \{\log(g+1) + 1\}^{-c}}{(g+1)^{a+(b+1)}} \leq \max\{1, (\log 2 + 1)^{-c}\} g^b. \tag{A.7}$$

For $g \geq 1$, let $\gamma = 2$ for $a > 0$ and $\gamma = c$ for $a = 0$. Further let

$$f_2(g) = (g+1)^{-a} \left(\frac{g}{g+1}\right)^b \{\log(g+1) + 1\}^{-c+\gamma},$$

which is bounded for $g \in (1, \infty)$ since $f_1(1) < \infty$ and $f_1(\infty) < \infty$. Then, for $g \geq 1$, we have

$$\frac{\pi(g; a, b, c)}{g+1} = \frac{f_2(g)}{(g+1)\{\log(g+1)+1\}^\gamma} \leq \frac{\max_{g \geq 1} f_2(g)}{(g+1)\{\log(g+1)+1\}^\gamma}. \quad (A.8)$$

By (A.7) and (A.8), we have

$$\int_0^\infty \frac{\pi(g; a, b, c)}{g+1} dg$$

$$\leq \max\{1, (\log 2 + 1)^{-c}\} \int_0^1 g^b dg + \int_1^\infty \frac{\max_{g \geq 1} f_2(g)}{(g+1)\{\log(g+1)+1\}^\gamma} dg$$

$$= \frac{\max\{1, (\log 2 + 1)^{-c}\}}{b+1} + \frac{(\log 2 + 1)^{1-\gamma}}{\gamma - 1},$$

which completes the proof of Part 2.

[Part 3] Let

$$f_3(g) = \frac{\{\log(g+1)+1\}^{1+c}}{(g+1)^{-a}} \left(\frac{g+1}{g}\right)^{b+1},$$

which is positive and bounded away from 0, for either $a > 0$ or $\{a = 0$ and $c \geq -1\}$. Then

$$\int_1^\infty \frac{dg}{g\pi(g; a, b, c)} = \int_1^\infty \frac{f_3(g)dg}{(g+1)\{\log(g+1)+1\}}$$

$$\geq \min_{g \geq 1} f_3(g) \int_1^\infty \frac{(g+1)^{-1}dg}{\log(g+1)+1} = \min_{g \geq 1} f_3(g) \left[\log(\{\log(g+1)+1\})\right]_1^\infty = \infty,$$

which completes the proof of Part 3. □

A.4 Proof of Theorem 2.5

For the proof of Theorem 2.5, recall $a = 0$. Hence it is convenient to use the notation

$$\xi(g) = \pi(g) = \left(\frac{g}{g+1}\right)^b \frac{1}{\{\log(g+1)+1\}^c}.$$

A.4.1 The Sequence k_i

Let $L(g) = \log(g+1) + 1$. Then the non-integrability

$$\int_0^\infty \frac{dg}{(g+1)L(g)} = \int_0^\infty \frac{dz}{z+1} = \infty$$

follows. Let

$$k_i(g) = \begin{cases} 1 - \dfrac{\log(L(g))}{\log(L(i))} & 0 < g < i \\ 0 & g \geq i. \end{cases}$$

For fixed g, $k_i(g)$ is increasing in i and $\lim_{i \to \infty} k_i(g) = 1$. Since $k_i(g) = 0$ for $g \geq i$, $\int_0^\infty \pi(g) k_i^2(g) dg < \infty$ even if $\int_0^\infty \pi(g) dg = \infty$. Note $k_i(g)$ is piecewise differentiable as

$$k_i'(g) = \begin{cases} -\dfrac{1}{(g+1)L(g)\log(L(i))} & 0 < g < i \\ 0 & g \geq i. \end{cases}$$

Then $k_i^2(g)$ is continuously differentiable since $\{k_i^2(g)\}' = 2k_i(g)k_i'(g)$ and $k_i(i) = 0$. Further we have

$$\sup_i |k_i'(g)| \{(g+1)L(g)\} \leq \begin{cases} 1/\log(L(1)) & 0 < g < 1 \\ 1/\log(L(g)) & g \geq 1, \end{cases}$$

and hence

$$\sup_i \{k_i'(g)\}^2 (g+1)L(g) \leq \begin{cases} 1/\{\log(L(1))\}^2 & 0 < g < 1 \\ \dfrac{1}{(g+1)L(g)\{\log(L(g))\}^2} & g \geq 1, \end{cases} \qquad (A.9)$$

which will be used in Sect. A.4.4.

A.4.2 Re-Expression of the Risk Difference

Let $v = w/2 = \|x\|^2/2$ and

$$\mathcal{F}(g; v) = \frac{1}{(2\pi)^{p/2}} \left(\exp\left(-\frac{v}{g+1} \right) - \exp(-v) \right),$$

which, for $b > -1$, gives

$$\lim_{g \to 0} \mathcal{F}(g; v) \left(\frac{g}{g+1} \right)^b = 0$$

as in (2.38). Let

$$v(g) = \frac{\xi(g)}{(g+1)^{p/2}}.$$

(A.10)

Then an integration by parts for $\hat{\mu}_i$ given by (2.18) gives

$$v \int_0^\infty \frac{(g+1)^{-p/2-1}}{(2\pi)^{p/2}} \exp\left(-\frac{v}{g+1}\right) \xi(g)k_i^2(g)dg$$

(A.11)

$$= \left[\frac{\mathcal{F}(g;v)\xi(g)k_i^2(g)}{(g+1)^{p/2-1}}\right]_0^\infty - \int_0^\infty \frac{\mathcal{F}(g;v)}{(g+1)^{p/2-1}}\left\{-\frac{p/2-1}{g+1} + \frac{d}{dg}\{\xi(g)k_i^2(g)\}\right\}dg$$

$$= (p/2-1)m_i(2v) - \frac{p/2-1}{(2\pi)^{p/2}}\int_0^\infty \frac{v(g)k_i^2(g)}{\exp(v)}dg$$

$$- b\int_0^\infty \frac{v(g)k_i^2(g)}{g}\mathcal{F}(g;v)dg + c\int_0^\infty \frac{v(g)k_i^2(g)}{L(g)}\mathcal{F}(g;v)dg$$

$$- 2\int_0^\infty (g+1)v(g)k_i(g)k_i'(g)\mathcal{F}(g;v)dg.$$

For $\hat{\mu}_\pi$, let $k_i \equiv 1$ in (A.11). Then $\|\hat{\mu}_\pi - \hat{\mu}_i\|^2 m_i(2v)$ given in (2.19) is

$$\|\hat{\mu}_\pi - \hat{\mu}_i\|^2 m_i(2v) = 4\frac{m_i(2v)}{2v}\{A_1(v) + 2A_2(v) + bA_3(v) - cA_4(v)\}^2,$$

where

$$A_1(v) = \frac{p/2-1}{\exp(v)}\left(\frac{\int_0^\infty v(g)k_i^2(g)dg}{(2\pi)^{p/2}m_i(2v)} - \frac{\int_0^\infty v(g)dg}{(2\pi)^{p/2}m_\pi(2v)}\right)$$

$$- \frac{c}{\exp(v)}\left(\frac{\int_0^\infty \{v(g)/L(g)\}k_i^2(g)dg}{(2\pi)^{p/2}m_i(2v)} - \frac{\int_0^\infty \{v(g)/L(g)\}dg}{(2\pi)^{p/2}m_\pi(2v)}\right)$$

$$+ b\left(\frac{\int_0^{1/2} g^{-1}\mathcal{F}(g;v)v(g)k_i^2(g)dg}{m_i(2v)} - \frac{\int_0^{1/2} g^{-1}\mathcal{F}(g;v)v(g)dg}{m_\pi(2v)}\right),$$

$$A_2(v) = \frac{\int_0^\infty (g+1)v(g)k_i(g)k_i'(g)\mathcal{F}(g;v)dg}{m_i(2v)},$$

$$A_3(v) = \frac{\int_{1/2}^\infty v(g)k_i^2(g)g^{-1}\mathcal{F}(g;v)dg}{m_i(2v)} - \frac{\int_{1/2}^\infty v(g)g^{-1}\mathcal{F}(g;v)dg}{m_\pi(2v)},$$

$$A_4(v) = \frac{\int_0^\infty \frac{v(g)k_i^2(g)dg}{L(g)\exp(v/(g+1))}}{(2\pi)^{p/2}m_i(2v)} - \frac{\int_0^\infty \frac{v(g)dg}{L(g)\exp(v/(g+1))}}{(2\pi)^{p/2}m_\pi(2v)}.$$

Further, by the inequality (Part 3 of Lemma A.3), we have

$$\|\hat{\mu}_\pi - \hat{\mu}_i\|^2 m_i(2v) \le 16 \frac{m_i(2v)}{2v} \{A_1^2(v) + 4A_2^2(v) + b^2 A_3^2(v) + c^2 A_4^2(v)\}.$$

As noted earlier, the proof is completed by proving dominated convergence for each of these 4 terms.

A.4.3 Dominated Convergence for the Term Involving A_1

Since $m_i(2v) \le m_\pi(2v)$ and $0 \le k_i^2 \le 1$, we have

$$|A_1(v)| \le \frac{p-2}{(2\pi)^{p/2}} \frac{\int_0^\infty v(g)dg}{\exp(v)m_i(2v)} + \frac{2|c|}{(2\pi)^{p/2}} \frac{\int_0^\infty \{v(g)/L(g)\}dg}{\exp(v)m_i(2v)}$$
$$+ 2|b| \frac{\int_0^{1/2} v(g)g^{-1}\mathcal{F}(g;v)dg}{m_i(2v)}.$$

For $g \in (0, 1/2)$, we have

$$(2\pi)^{p/2}\frac{\mathcal{F}(g;v)}{g} = \exp\left(\frac{-v}{g+1}\right)\frac{1}{g}\left\{1 - \exp\left(\frac{-gv}{g+1}\right)\right\} \le \exp\left(-\frac{2v}{3}\right)\frac{v}{g+1}$$
$$\le v\exp\left(-\frac{2v}{3}\right),$$

where the first inequality follows from the general inequality, $1 - e^{-y} \le y$ for $y \in \mathbb{R}$. Further by Part 7 of Lemma A.5, we have

$$v\exp\left(-\frac{2v}{3}\right) = \left\{v\exp\left(-\frac{2v}{21}\right)\right\}\exp\left(-\frac{4v}{7}\right) \le \frac{21}{2}\exp\left(-\frac{4v}{7}\right),$$

and hence

$A_1(v)$
$$\le \frac{(p-2)\int_0^\infty v(g)dg}{\exp(v)(2\pi)^{p/2}m_i(2v)} + \frac{2|c|\int_0^\infty \{v(g)/L(g)\}dg}{\exp(v)(2\pi)^{p/2}m_i(2v)} + \frac{21|b|\int_0^{1/2} v(g)dg}{\exp(4v/7)(2\pi)^{p/2}m_i(2v)}$$
$$\le \frac{\mathcal{A}_1}{\exp(4v/7)(2\pi)^{p/2}m_i(2v)},$$

where

$$\mathcal{A}_1 = (p-2)\int_0^\infty v(g)dg + 2|c|\int_0^\infty \frac{v(g)}{L(g)}dg + 21|b|\int_0^{1/2} v(g)dg.$$

Note $m_1(2v) \le m_i(2v)$ for $i \in \mathbb{N}$ and

$$\exp(v)(2\pi)^{p/2} m_1(2v) = \int_0^\infty v(g)k_1^2(g)\exp\left(\frac{gv}{g+1}\right)dg \geq \int_0^\infty v(g)k_1^2(g)dg.$$

Hence

$$m_i(2v)A_1^2(v) \leq \frac{\mathcal{A}_1^2\exp(-8v/7)}{(2\pi)^p m_i(2v)} \leq \frac{1}{(2\pi)^{p/2}}\frac{\mathcal{A}_1^2\exp(-v/7)}{\exp(v)(2\pi)^{p/2}m_1(2v)}$$

$$\leq \frac{1}{(2\pi)^{p/2}}\frac{\mathcal{A}_1^2\exp(-v/7)}{\int_0^\infty v(g)k_1^2(g)dg}.$$

Recall $v = w/2 = \|x\|^2/2$. By Part 2 of Lemma A.2, we have

$$\frac{1}{(2\pi)^{p/2}}\int_{\mathbb{R}^p}\frac{1}{\|x\|^2}\exp\left(-\frac{\|x\|^2}{2\beta}\right)dx = \frac{\beta^{p/2-1}}{p-2}. \tag{A.12}$$

Then we have

$$\int_{\mathbb{R}^p}\frac{m_i(\|x\|^2)}{\|x\|^2}A_1^2(\|x\|^2/2)dx \leq \frac{\mathcal{A}_1^2}{\int_0^\infty v(g)k_1^2(g)dg}\int_{\mathbb{R}^p}\frac{\exp(-\|x\|^2/14)dx}{(2\pi)^{p/2}\|x\|^2}$$

$$= \frac{7^{p/2-1}\mathcal{A}_1^2}{(p-2)\int_0^\infty v(g)k_1^2(g)dg} < \infty.$$

A.4.4 Dominated Convergence for the Term Involving A_2

Note $0 \leq \mathcal{F}(g;v) \leq (2\pi)^{-p/2}\exp(-v/(g+1))$ and hence

$$|A_2(v)| \leq \frac{\int_0^\infty (g+1)v(g)k_i(g)|k_i'(g)|\exp(-v/(g+1))dg}{(2\pi)^{p/2}m_i(2v)}.$$

The Cauchy-Schwarz inequality (Part 1 of Lemma A.3) gives

$$\left(\int_0^\infty\frac{(g+1)v(g)k_i(g)|k_i'(g)|}{\exp(v/(g+1))}dg\right)^2$$

$$\leq (2\pi)^{p/2}m_i(2v)\int_0^\infty\frac{(g+1)^2v(g)\{k_i'(g)\}^2}{\exp(v/(g+1))}dg.$$

Recall $v(g) = \xi(g)/(g+1)^{p/2}$ as in (A.10). Then, we have

$$m_i(2v)A_2^2(v) \le \frac{1}{(2\pi)^{p/2}} \int_0^\infty \frac{(g+1)^2 \xi(g)\{k_i'(g)\}^2}{(g+1)^{p/2} \exp(v/(g+1))} dg$$

$$\le \frac{1}{(2\pi)^{p/2}} \int_0^\infty \frac{(g+1)^2 L(g)\{k_i'(g)\}^2}{(g+1)^{p/2} \exp(v/(g+1))} \left(\frac{g}{g+1}\right)^b dg,$$

where the second inequality follows from the fact

$$\xi(g) = \left(\frac{g}{g+1}\right)^b \frac{1}{\{\log(g+1)+1\}^c} \le \left(\frac{g}{g+1}\right)^b L(g)$$

for $|c| \le 1$. Then, by (A.12), we have

$$\int_{\mathbb{R}^p} \frac{m_i(\|x\|^2)}{\|x\|^2} A_2^2(\|x\|^2/2) dx \le \int_0^\infty \frac{g+1}{p-2} \left(\frac{g}{g+1}\right)^b L(g) \sup_i \{k_i'(g)\}^2 dg.$$

By Part 3 of Lemma A.5 and (A.9), we have

$$(p-2) \int_{\mathbb{R}^p} \frac{m_i(\|x\|^2)}{\|x\|^2} A_2^2(\|x\|^2/2) dx \le \int_0^\infty (g+1) \left(\frac{g}{g+1}\right)^b L(g) \sup_i \{k_i'(g)\}^2 dg$$

$$\le \int_0^1 \frac{g^b dg}{\{\log L(1)\}^2} + 2 \int_1^\infty \frac{\{\log(L(g))\}^{-2} dg}{(g+1)L(g)}$$

$$= \frac{(b+1)^{-1}}{\{\log L(1)\}^2} + \frac{2}{\log L(1)} < \infty. \qquad (A.13)$$

A.4.5 Dominated Convergence for the Term Involving A_3

For $g \in (1/2, \infty), \mathcal{F}(g; v) \le (2\pi)^{-p/2} \exp(-v/(g+1)), g \ge (g+1)/3$ and hence

$$(2\pi)^{p/2} \frac{\mathcal{F}(g; v)}{g} I_{(1/2,\infty)}(g) \le \frac{3}{g+1} \exp\left(\frac{-v}{g+1}\right).$$

Then

$$(2\pi)^{p/2} m_i(2v)A_3^2(v) \le 18 \left(\frac{\{\int_0^\infty (g+1)^{-1} v(g)k_i^2(g) \exp(-v/(g+1)) dg\}^2}{m_i(2v)} \right.$$

$$\left. + \frac{\{\int_0^\infty (g+1)^{-1} v(g) \exp(-v/(g+1)) dg\}^2}{m_\pi(2v)} \right)$$

$$\le \int_0^\infty \frac{36 v(g) dg}{(g+1)^2 \exp(v/(g+1))}, \qquad (A.14)$$

where the second inequality follows from the Cauchy–Schwarz inequality (Part 1 of Lemma A.3). Recall $v(g) = \xi(g)/(g+1)^{p/2}$ as in (A.10). By (A.12) and (A.14), we have

$$\int_{\mathbb{R}^p} \frac{m_i(\|x\|^2)}{\|x\|^2} A_3^2(\|x\|^2/2)dx \le \frac{36}{p-2} \int_0^\infty \frac{\xi(g)}{(g+1)^3}dg < \infty.$$

A.4.6 Dominated Convergence for the Term Involving A_4

We rewrite $\{(2\pi)^{p/2} m_i(w)\}^2 A_4^2(v)$ as

$$\{(2\pi)^{p/2} m_i(w)\}^2 A_4^2(v)$$
$$= \left(\int_0^\infty \frac{v(g)k_i^2(g)}{\exp(v/(g+1))} \left\{ \frac{1}{L(g)} - \frac{1}{m_\pi(2v)} \int_0^\infty \frac{v(g)\{L(g)\}^{-1}dg}{\exp(v/(g+1))} \right\} dg \right)^2.$$

By the Cauchy-Schwarz inequality (Part 1 of Lemma A.3), we have

$$(2\pi)^{p/2} m_i(2v) A_4^2(v) \tag{A.15}$$
$$\le \int_0^\infty \frac{v(g)k_i^2(g)}{\exp(v/(g+1))} \left\{ \frac{1}{L(g)} - \frac{\int_0^\infty \{L(g)\}^{-1}v(g)\exp(-v/(g+1))dg}{m_\pi(2v)} \right\}^2 dg$$
$$\le \int_0^\infty \frac{v(g)}{\exp(v/(g+1))} \left\{ \frac{1}{L(g)} - \frac{\int_0^\infty \{L(g)\}^{-1}v(g)\exp(-v/(g+1))dg}{m_\pi(2v)} \right\}^2 dg$$
$$= \int_0^\infty \frac{v(g)dg}{\exp(v/(g+1))L^2(g)} - \frac{\{\int_0^\infty \{L(g)\}^{-1}v(g)\exp(-v/(g+1))dg\}^2}{\int_0^\infty v(g)\exp(-v/(g+1))dg},$$

where the second inequality follows from the fact $k_i^2(g) \le 1$.
When $-1 < c \le 1$, we have

$$(2\pi)^{p/2} m_i(w) A_4^2(v) \le \int_0^\infty \frac{v(g)dg}{\exp(v/(g+1))L^2(g)}$$
$$= \int_0^\infty \left(\frac{g}{g+1} \right)^b \frac{\exp(-v/(g+1))}{(g+1)^{p/2}L^{2+c}(g)} dg$$

and, by (A.12) and Part 3 of Lemma A.5,

$$(p-2) \int_{\mathbb{R}^p} \frac{m_i(\|x\|^2)}{\|x\|^2} A_4^2(\|x\|^2/2)dx \le \int_0^\infty \left(\frac{g}{g+1} \right)^b \frac{dg}{(g+1)L^{2+c}(g)}$$
$$\le \int_0^1 \frac{(g+1)^{-1}g^b dg}{L^{2+c}(g)} + 2\int_0^\infty \frac{(g+1)^{-1}dg}{L^{2+c}(g)} \le \frac{1}{b+1} + \frac{2}{1+c} < \infty,$$

which completes the proof for the case $-1 < c \leq 1$.

When $c = -1$, a more careful treatment is needed. In (A.15) we have

$$
\int_0^\infty \frac{v(g)dg}{L(g)\exp(v/(g+1))}
$$

$$
= \int_0^\infty (g+1)^{-p/2} \exp\left(-\frac{v}{g+1}\right)\left(1 - \frac{1}{g+1}\right)^b dg
$$

$$
\geq \int_0^\infty (g+1)^{-p/2} \exp\left(-\frac{v}{g+1}\right)\left(1 - \frac{\max(b,1)}{g+1}\right) dg
$$

$$
= v^{-p/2+1} \int_0^v \frac{t^{p/2-2}}{\exp(t)}\left(1 - t\frac{\max(b,1)}{v}\right) dt
$$

$$
\geq v^{-p/2+1}\left(\Gamma(p/2-1) - \int_v^\infty \frac{t^{p/2-2}}{\exp(t)}dt - \max(b,1)\frac{\Gamma(p/2)}{v}\right),
$$

where the first inequality follows from Part 1 of Lemma A.5. Hence there exist $Q_1 > 0$ and $v_1 > \exp(2Q_1)$ such that

$$
\int_0^\infty \frac{v(g)dg}{L(g)\exp(v/(g+1))} \geq \frac{\Gamma(p/2-1)}{v^{p/2-1}}\left(1 - \frac{Q_1}{L(v)}\right) \geq \frac{1}{2}\frac{\Gamma(p/2-1)}{v^{p/2-1}}, \quad \text{(A.16)}
$$

for all $v \geq v_1$. Further we have

$$
\int_0^\infty \frac{v(g)dg}{\exp(v/(g+1))} \tag{A.17}
$$

$$
= \int_0^\infty (g+1)^{-p/2} \exp\left(-\frac{v}{g+1}\right)\left(1 - \frac{1}{g+1}\right)^b \{\log(g+1) + 1\}dg
$$

$$
= \int_0^\infty (g+1)^{-p/2} \exp\left(-\frac{v}{g+1}\right)\left(1 - \frac{1}{g+1}\right)^b \{\log v - \log\{v/(g+1)\} + 1\}dg
$$

$$
= L(v-1)\int_0^\infty \frac{v(g)dg}{L(g)\exp(v/(g+1))} - \frac{1}{v^{p/2-1}}\int_0^v \frac{(\log t)t^{p/2-2}}{\exp(t)}(1-t/v)^b dt
$$

$$
\leq L(v)\int_0^\infty \frac{v(g)dg}{L(g)\exp(v/(g+1))} + \frac{\max(1,\{1-1/v_1\}^b)}{v^{p/2-1}}\int_0^1 \frac{|\log t|t^{p/2-2}}{\exp(t)}dt,
$$

where the last inequality follows under all $v \geq v_1$. By (A.16) and (A.17), for all $v \geq v_1$, we have

$$
\frac{\int_0^\infty v(g)\exp(-v/(g+1))dg}{\int_0^\infty \{v(g)/L(g)\}\exp(-v/(g+1))dg} \leq L(v) + Q_2 \tag{A.18}
$$

where

$$
Q_2 = 2\max\left(1,\{1-1/v_1\}^b\right)\frac{\int_0^1 |\log t|t^{p/2-2}\exp(-t)dt}{\Gamma(p/2-1)}.
$$

Further, for all $v \geq \max(v_1, \exp(Q_2))$, we have $Q_2/L(v) < 1$ and hence

$$L(v) + Q_2 = L(v)\left\{1 + \frac{Q_2}{L(v)}\right\} \leq \frac{L(v)}{1 - Q_2/L(v)}. \tag{A.19}$$

Then, by (A.16), (A.18) and (A.19), we have

$$\frac{\{\int_0^\infty \{v(g)/L(g)\} \exp(-v/(g+1))dg\}^2}{\int_0^\infty v(g) \exp(-v/(g+1))dg} \tag{A.20}$$

$$\geq \frac{\Gamma(p/2 - 1)}{v^{p/2-1}L(v)}\left(1 - \frac{Q_1 + Q_2}{L(v)}\right),$$

for all $v \geq \max(v_1, \exp(Q_2))$. Let $v_2 = \max(v_1, \exp(Q_2))$ and $Q_3 = Q_1 + Q_2$. Then, by (A.20), we have

$$\frac{1}{(2\pi)^{p/2}} \int_{\mathbb{R}^p} \frac{\{\int_0^\infty \{v(g)/L(g)\} \exp(-\|x\|^2/\{2(g+1)\})dg\}^2}{\|x\|^2 \int_0^\infty v(g) \exp(-\|x\|^2/\{2(g+1)\})dg} dx$$

$$\geq \frac{1}{(2\pi)^{p/2}} \int_{\|x\|^2 > 2v_2} \frac{1}{\|x\|^2} \frac{2^{p/2-1}\Gamma(p/2-1)}{(\|x\|^2)^{p/2-1}L(\|x\|^2/2)}\left(1 - \frac{Q_3}{L(\|x\|^2/2)}\right)dx$$

$$= \frac{1}{p-2}\left(\int_{v_2}^\infty \frac{dg}{gL(g)} - \int_{v_2}^\infty \frac{Q_3 dg}{g\{L(g)\}^2}\right). \tag{A.21}$$

By (A.12), (A.15) and (A.21), we have

$$(p-2) \int_{\mathbb{R}^p} \frac{m_i(\|x\|^2)}{\|x\|^2} A_4^2(\|x\|^2/2)dx$$

$$\leq \int_0^\infty \frac{dg}{(g+1)L(g)} - \int_{v_2}^\infty \frac{dg}{gL(g)} + \int_{v_2}^\infty \frac{Q_3 dg}{g\{L(g)\}^2}$$

$$= \int_0^{v_2} \frac{dg}{(g+1)L(g)} - \int_{v_2}^\infty \frac{dg}{g(g+1)L(g)} + \int_{v_2}^\infty \frac{Q_3 dg}{g\{L(g)\}^2}$$

$$= \int_0^{v_2} \frac{dg}{(g+1)L(g)} + \int_{v_2}^\infty \frac{Q_3 dg}{g\{L(g)\}^2} < \infty.$$

We conclude then, by dominated convergence, that $\Delta_i \to 0$, which completes the proof.

A.5 Properties of $\varphi(g; w)$

This section gives properties of the function $\varphi(g; w)$ used in the proof of Theorem 3.5 in Sect. A.6.

Lemma A.9 1. $w \dfrac{(n/2 + 1 - a)(g + 1)^{-p/2-a-1}}{\{1 + w/(g + 1)\}^{p/2+n/2+1}}$

$$= \begin{cases} \dfrac{d}{dg}\left\{\left(1 - \dfrac{w}{g + 1 + w}\right)^{n/2+1-a}\right\}(g + 1 + w)^{-p/2+1-a} \\ \dfrac{d}{dg}\left\{\left(1 - \dfrac{w}{g + 1 + w}\right)^{n/2+1-a}\varphi(g; w)\right\}(g + 1 + w)^{-p/2+1-a}, \end{cases}$$

where

$$\varphi(g; w) = 1 - \left(\frac{g + 1 + w}{(g + 1)(w + 1)}\right)^{n/2+1-a}. \tag{A.22}$$

2. Let $n \geq 2$, $b > -1$, $-p/2 + 1 < a \leq n/2$ in $\varphi(g; w)$ given by (A.22). Then $\varphi(g; w)$ satisfies

$$\varphi(g; w) \leq \max(1, n/2 + 1 - a)\frac{g}{g + 1} \quad and \quad 1 - \varphi(g; w) \leq \frac{1}{g + 1} + \frac{1}{w^{1/8}}.$$

Proof [Part 1]

$w \dfrac{(n/2 + 1 - a)(g + 1)^{-p/2-a-1}}{\{1 + w/(g + 1)\}^{p/2+n/2+1}}$

$= w(n/2 + 1 - a)\left(1 - \dfrac{w}{g + 1 + w}\right)^{n/2-a}(g + 1 + w)^{-p/2-a-1}$

$$= \begin{cases} \dfrac{d}{dg}\left\{\left(1 - \dfrac{w}{g + 1 + w}\right)^{n/2+1-a}\right\}(g + 1 + w)^{-p/2+1-a} \\ \dfrac{d}{dg}\left\{\left(1 - \dfrac{w}{g + 1 + w}\right)^{n/2+1-a} - \left(1 - \dfrac{w}{1 + w}\right)^{n/2+1-a}\right\}(g + 1 + w)^{-p/2+1-a}, \end{cases}$$

where

$$\left(1 - \frac{w}{g + 1 + w}\right)^{n/2+1-a} - \left(1 - \frac{w}{1 + w}\right)^{n/2+1-a} = \left(1 - \frac{w}{g + 1 + w}\right)^{n/2+1-a}\varphi(g; w).$$

[Part 2] By Part 1 of Lemma A.5, $(1 - x)^{n/2+1-a} \geq 1 - \max(n/2 + 1 - a, 1)x$ and

$$\varphi(g; w) \leq \max(n/2 + 1 - a, 1)\frac{w}{w + 1}\frac{g}{g + 1} \leq \max(n/2 + 1 - a, 1)\frac{g}{g + 1}$$

follows. Further, $1 - \varphi(g; w)$ for $a \leq n/2$ is bounded as follows:

$$1 - \varphi(g; w) \le \left(\frac{g + 1 + w}{(g + 1)(w + 1)}\right)^{n/2 + 1 - a} \le \frac{g + 1 + w}{(g + 1)(w + 1)}$$

$$\le \frac{1}{g + 1} + \frac{1}{w + 1} \le \frac{1}{g + 1} + \frac{1}{w^{1/8}}. \qquad \square$$

A.6 Proof of Theorem 3.5

By the assumption of the theorem, the measure $\Pi(dg)$ in (3.27) has the density $\xi(g) = \{g/(g + 1)\}^b \{\log(g + 1) + 1\}^{-c}$. Then the Bayes risk difference (3.33) between $\hat{\theta}_\pi$ given by (3.31) with $\xi(g)$ above and $\hat{\theta}_{\pi i}$ under $\bar{\pi}_i$ given by (3.32) with $\xi(g)$ above and with

$$k_i(g) = \begin{cases} 1 - \dfrac{\log(\log(g + 1) + 1)}{\log(\log(i + 1) + 1)} & 0 < g < i \\ 0 & g \ge i, \end{cases}$$

is

$$\tilde{r}(\hat{\theta}_\pi; \bar{\pi}_i) - \tilde{r}(\hat{\theta}_{\pi i}; \bar{\pi}_i) = \int_{\mathbb{R}^p} \{\psi_\pi(\|z\|^2) - \psi_{\pi i}(\|z\|^2)\}^2 \|z\|^2 M_1(z, \pi_i) dz,$$

where the integrand is

$$\{\psi_\pi(\|z\|^2) - \psi_{\pi i}(\|z\|^2)\}^2 \|z\|^2 M_1(z, \pi_i) \tag{A.23}$$

$$= w\left(\frac{\int_0^\infty (g + 1)^{-p/2 - 1}\{1 + w/(g + 1)\}^{-(p+n)/2 - 1}\xi(g)dg}{\int_0^\infty (g + 1)^{-p/2}\{1 + w/(g + 1)\}^{-(p+n)/2 - 1}\xi(g)dg}\right.$$

$$\left. - \frac{\int_0^\infty (g + 1)^{-p/2 - 1}\{1 + w/(g + 1)\}^{-(p+n)/2 - 1}k_i^2(g)\xi(g)dg}{\int_0^\infty (g + 1)^{-p/2}\{1 + w/(g + 1)\}^{-(p+n)/2 - 1}k_i^2(g)\xi(g)dg}\right)^2$$

$$\times \frac{\Gamma((p + n)/2 + 1)2^{(p+n)/2 + 1}}{q_1(p, n)} \int_0^\infty \frac{(g + 1)^{-p/2}k_i^2(g)\xi(g)dg}{\{1 + w/(g + 1)\}^{(p+n)/2 + 1}},$$

where $w = \|z\|^2$.

Part 1 of Lemma A.9 with $a = 0$ gives

$$w\frac{(n/2 + 1)(g + 1)^{-p/2 - 1}}{\{1 + w/(g + 1)\}^{p/2 + n/2 + 1}} \tag{A.24}$$

$$= \frac{d}{dg}\left\{\left(1 - \frac{w}{g + 1 + w}\right)^{n/2 + 1} - \left(1 - \frac{w}{1 + w}\right)^{n/2 + 1}\right\}(g + 1 + w)^{-p/2 + 1}$$

$$= \frac{d}{dg}\left\{\left(1 - \frac{w}{g + 1 + w}\right)^{n/2 + 1}\varphi(g; w)\right\}(g + 1 + w)^{-p/2 + 1},$$

where

$$\varphi(g; w) = 1 - \left(\frac{g + 1 + w}{(g + 1)(w + 1)}\right)^{n/2 + 1}. \tag{A.25}$$

By (A.24), an integration by parts gives

$$(n/2+1)w \int_0^\infty \frac{k_i^2(g)\{g/(g+1)\}^b \{\log(g+1)+1\}^{-c}}{(g+1)^{p/2+1}\{1+w/(g+1)\}^{(p+n)/2+1}} dg \qquad (A.26)$$

$$= (p/2-1) \int_0^\infty \frac{k_i^2(g)\{g/(g+1)\}^b \{\log(g+1)+1\}^{-c}}{(g+1)^{p/2}\{1+w/(g+1)\}^{(p+n)/2+1}} dg$$

$$- \int_0^\infty \tilde{\varphi}(g;w) \frac{k_i^2(g)\{g/(g+1)\}^b \{\log(g+1)+1\}^{-c}}{(g+1)^{p/2}\{1+w/(g+1)\}^{(p+n)/2+1}} dg$$

$$+ c \int_0^\infty \varphi(g;w) \frac{1+w/(g+1)}{\log(g+1)+1} \frac{k_i^2(g)\{g/(g+1)\}^b \{\log(g+1)+1\}^{-c}}{(g+1)^{p/2}\{1+w/(g+1)\}^{(p+n)/2+1}} dg$$

$$- \int_0^\infty \varphi(g;w) \frac{2k_i(g)k_i'(g)(g+1+w)\{g/(g+1)\}^b \{\log(g+1)+1\}^{-c}}{(g+1)^{p/2}\{1+w/(g+1)\}^{(p+n)/2+1}} dg,$$

where

$$\tilde{\varphi}(g;w) = b \frac{g+1+w}{g+1} \frac{\varphi(g;w)}{g} + (p/2-1)\{1-\varphi(g;w)\}. \qquad (A.27)$$

Similarly, by (A.24), we have

$$(n/2+1)w \int_0^\infty \frac{\{g/(g+1)\}^b \{\log(g+1)+1\}^{-c}}{(g+1)^{p/2+1}\{1+w/(g+1)\}^{(p+n)/2+1}} dg \qquad (A.28)$$

$$= (p/2-1) \int_0^\infty \frac{\{g/(g+1)\}^b \{\log(g+1)+1\}^{-c}}{(g+1)^{p/2}\{1+w/(g+1)\}^{(p+n)/2+1}} dg$$

$$- \int_0^\infty \tilde{\varphi}(g;w) \frac{\{g/(g+1)\}^b \{\log(g+1)+1\}^{-c}}{(g+1)^{p/2}\{1+w/(g+1)\}^{(p+n)/2+1}} dg$$

$$+ c \int_0^\infty \varphi(g;w) \frac{1+w/(g+1)}{\log(g+1)+1} \frac{\{g/(g+1)\}^b \{\log(g+1)+1\}^{-c}}{(g+1)^{p/2}\{1+w/(g+1)\}^{(p+n)/2+1}} dg.$$

Let

$$\mathcal{F}(g;w) = \frac{\{g/(g+1)\}^b \{\log(g+1)+1\}^{-c}}{(g+1)^{p/2}\{1+w/(g+1)\}^{(p+n)/2+1}}.$$

Then, by (A.26) and (A.28), the multiple of the integrand given by (A.23) is

$$\frac{(n/2+1)^{-2}q_1(p,n)}{\Gamma((p+n)/2+1)2^{(p+n)/2+1}}\{\psi_\pi(\|z\|^2)-\psi_{\pi i}(\|z\|^2)\}^2\|z\|^2 M_1(z,\pi_i)$$

$$=\frac{1}{w}\left(\frac{\int_0^\infty \tilde{\varphi}(g;w)k_i^2(g)\mathcal{F}(g;w)dg}{\int_0^\infty k_i^2(g)\mathcal{F}(g;w)dg}-\frac{\int_0^\infty \tilde{\varphi}(g;w)\mathcal{F}(g;w)dg}{\int_0^\infty \mathcal{F}(g;w)dg}\right.$$

$$-c\frac{\int_0^\infty\{1+w/(g+1)\}\{\log(g+1)+1\}^{-1}k_i^2(g)\mathcal{F}(g;w)\varphi(g;w)dg}{\int_0^\infty k_i^2(g)\mathcal{F}(g;w)dg}$$

$$+c\frac{\int_0^\infty\{1+w/(g+1)\}\{\log(g+1)+1\}^{-1}\mathcal{F}(g;w)\varphi(g;w)dg}{\int_0^\infty \mathcal{F}(g;w)dg}$$

$$\left.+2\frac{\int_0^\infty\{k_i'(g)/k_i(g)\}(g+1+w)\mathcal{F}(g;w)\varphi(g;w)dg}{\int_0^\infty k_i^2(g)\mathcal{F}(g;w)dg}\right)^2\times\int_0^\infty k_i^2(g)\mathcal{F}(g;w)dg.$$

Note $0\le\varphi(g;w)\le 1$. Further, by the Cauchy-Schwarz inequality (Parts 3 and 1 of Lemma A.3), we have

$$\frac{(n/2+1)^{-2}q_1(p,n)}{\Gamma((p+n)/2+1)2^{(p+n)/2+1}}\{\psi_\pi(\|z\|^2)-\psi_{\pi i}(\|z\|^2)\}^2\|z\|^2 M_1(z,\pi_i)$$

$$\le\frac{5}{w}\int_0^\infty\left\{2\tilde{\varphi}^2(g;w)+2c^2\left(\frac{1+w/(g+1)}{\log(g+1)+1}\right)^2+4\{k_i'(g)\}^2(g+1+w)^2\right\}\mathcal{F}(g;w)dg.$$

For $\tilde{\varphi}^2(g;w)$, by Part 3 of Lemma A.3 and Part 2 of Lemma A.9, we have

$$\tilde{\varphi}^2(g;w)\le 3\left(\frac{\{|b|(n/2+1)+(p/2-1)\}^2}{(g+1)^2}+\frac{|b|^2(n/2+1)^2 w^2}{(g+1)^4}+\frac{(p/2-1)^2}{w^{1/4}}\right).$$

By Part 3 of Lemma A.2, we have

$$\int_{\mathbb{R}^p}\frac{1}{\|z\|^{2+2\alpha}}\left(1+\frac{\|z\|^2}{g+1}\right)^{-p/2-n/2-1}dz=\frac{\pi^{p/2}}{\Gamma(p/2)}\frac{B(p/2-1-\alpha,n/2+2+\alpha)}{(g+1)^{-p/2+1+\alpha}},$$

and hence

$$\int_{\mathbb{R}^p}\frac{1}{\|z\|^{2+2\alpha}}\mathcal{F}(g;\|z\|^2)dz$$

$$=\frac{\pi^{p/2}}{\Gamma(p/2)}B(p/2-1-\alpha,n/2+2+\alpha)\frac{\{g/(g+1)\}^b\{\log(g+1)+1\}^{-c}}{(g+1)^{1+\alpha}}.$$

As in (A.13), the integral

$$\int_0^\infty (g+1)\sup_i\{k_i'(g)\}^2 \frac{\{g/(g+1)\}^b}{\{L(g)\}^c}\, dg,$$

for $b > -1$ and $c > -1$ is integrable. Further since all the integrals

$$\int_0^\infty \frac{\{g/(g+1)\}^b dg}{(g+1)^3\{L(g)\}^c}, \quad \int_0^\infty \frac{\{g/(g+1)\}^b dg}{(g+1)^{1+1/4}\{L(g)\}^c}, \quad \int_0^\infty \frac{\{g/(g+1)\}^b dg}{(g+1)\{L(g)\}^{2+c}},$$

for $b > -1$ and $c > -1$ are finite, it follows that

$$\int_{\mathbb{R}^p} \{\psi_\pi(\|z\|^2) - \psi_{\pi i}(\|z\|^2)\}^2 \|z\|^2 M_1(z, \pi_i)\, dz < \infty.$$

Then by the dominated convergence theorem, we have

$$\lim_{i\to\infty} \{\tilde{r}(\hat{\mu}_\pi; \pi_i) - \tilde{r}(\hat{\mu}_{\pi i}; \pi_i)\} = 0$$

which, through the Blyth method, implies the admissibility of $\hat{\mu}_\pi$ within the class of equivariant estimators, as was to be shown.

A.7 Lemmas Used in the Proof of Theorem 3.6

This section is devoted to showing that the integral of each term involving A_1, A_2, and A_3 respectively, in the proof of Theorem 3.6, approaches 0 as $i \to \infty$.

A.7.1 Proof for A_1

By the Cauchy-Schwarz inequality (Part 1 of Lemma A.3),

$$A_1(w, s; i) \tag{A.29}$$

$$\leq \iint \frac{F\pi dgd\eta}{(g+1)^2} \iint \left(\frac{1}{\iint F\pi dgd\eta} - \frac{h_i^2}{\iint Fh_i^2\pi dgd\eta}\right)^2 F\pi dgd\eta \iint Fh_i^2\pi dgd\eta.$$

Note

$$\left(\frac{1}{\iint F\pi dgd\eta} - \frac{h_i^2}{\iint Fh_i^2\pi dgd\eta}\right)^2 \tag{A.30}$$

$$= \left(\frac{1}{\sqrt{\iint F\pi dgd\eta}} - \frac{h_i}{\sqrt{\iint Fh_i^2\pi dgd\eta}}\right)^2\left(\frac{1}{\sqrt{\iint F\pi dgd\eta}} + \frac{h_i}{\sqrt{\iint Fh_i^2\pi dgd\eta}}\right)^2$$

$$= \frac{1}{\iint F\pi dgd\eta \iint Fh_i^2\pi dgd\eta}\left(1 - \frac{h_i\sqrt{\iint F\pi dgd\eta}}{\sqrt{\iint Fh_i^2\pi dgd\eta}}\right)^2\left(\frac{\sqrt{\iint Fh_i^2\pi dgd\eta}}{\sqrt{\iint F\pi dgd\eta}} + h_i\right)^2$$

$$\leq \frac{2^2}{\iint F\pi dgd\eta \iint Fh_i^2\pi dgd\eta}\left(1 - \frac{h_i\sqrt{\iint F\pi dgd\eta}}{\sqrt{\iint Fh_i^2\pi dgd\eta}}\right)^2,$$

where the inequality follows from the fact $0 \leq h_i \leq 1$.
Further,

$$\frac{1}{2\iint F\pi dgd\eta}\iint F\pi\left(1 - \frac{h_i\sqrt{\iint F\pi dgd\eta}}{\sqrt{\iint Fh_i^2\pi dgd\eta}}\right)^2 dgd\eta \tag{A.31}$$

$$= 1 - \sqrt{\frac{(\iint Fh_i\pi dgd\eta)^2}{\iint F\pi dgd\eta \iint Fh_i^2\pi dgd\eta}} \leq 1 - \frac{(\iint Fh_i\pi dgd\eta)^2}{\iint F\pi dgd\eta \iint Fh_i^2\pi dgd\eta},$$

where the inequality follows from the fact that

$$\frac{(\iint Fh_i\pi dgd\eta)^2}{\iint F\pi dgd\eta \iint Fh_i^2\pi dgd\eta} \in (0, 1),$$

which follows from the Cauchy–Schwarz inequality (Part 1 of Lemma A.3). By (A.29), (A.30) and (A.31), we have

$$A_1(w, s; i) \leq 8\tilde{A}_1(w, s; i)\iint \frac{F\pi(g)dgd\eta}{(g + 1)^2}, \tag{A.32}$$

where

$$\tilde{A}_1(w, s; i) = 1 - \frac{(\iint Fh_i\pi dgd\eta)^2}{\iint F\pi dgd\eta \iint Fh_i^2\pi dgd\eta}. \tag{A.33}$$

For (A.33), the following lemma is useful.

Lemma A.10 *Assume* $-p/2 + 1 < a < n/2 + 2$. *Then there exists a positive constant* q_2, *independent of* i, x *and* s, *such that*

$$\tilde{A}_1(w, s; i) \leq q_2(1 + |\log s|)^{-2}. \tag{A.34}$$

Proof See Sect. A.8.   □

By (A.32), (A.34) and Lemma A.11 below, we have

$$\iint \|x\|^2 s^{n/2-1} A_1(\|x\|^2/s, s; i) dx ds \leq 8q_2 \iiiint \frac{\|x\|^2 s^{n/2-1} F\pi(g) dg d\eta dx ds}{(g+1)^2 (1+|\log s|)^2}$$

$$\leq 16q_3(0) q_2 B(a, b+1) < \infty,$$

where $q_3(\beta)$ is given by

$$q_3(\beta) = 2^{p/2+n/2+1} \pi^{p/2} \frac{\Gamma(p/2+1-\beta)\Gamma(n/2+\beta)}{\Gamma(p/2)}. \tag{A.35}$$

For all w and s, $\lim_{i\to\infty} A_1(w, s; i) = 0$. Thus, by the dominated convergence theorem,

$$\lim_{i\to\infty} \iint \|x\|^2 s^{n/2-1} A_1(\|x\|^2/s, s; i) dx ds = 0,$$

which completes the proof for A_1.

Lemma A.11 *Assume* $-n/2 < \beta < p/2 + 1$. *Then*

$$\iiint \|x\|^2 s^{n/2-1} \frac{F(g, \eta; \|x\|^2/s, s)}{(1+|\log s|)^2} d\eta dx ds = 2q_3(\beta)(g+1),$$

where $q_3(\beta)$ *is given by* (A.35).

Proof

$$\iiint \|x\|^2 s^{n/2-1} \frac{F(g, \eta; \|x\|^2/s, s)}{(1+|\log s|)^2} d\eta dx ds$$

$$= \iiint \|x\|^2 s^{n/2-1} \frac{\eta^{p/2+n/2}}{(g+1)^{p/2}} \exp\left(-\frac{\eta s}{2}\left\{1 + \frac{\|x\|^2/s}{g+1}\right\}\right) \frac{d\eta dx ds}{(1+|\log s|)^2}$$

$$= \iiint \frac{s^{p/2+n/2-1} \|y\|^2 (1+g) s}{(1+g)^{-p/2}} \frac{\eta^{p/2+n/2}}{(g+1)^{p/2}} \exp\left(-\frac{\eta s}{2}\left\{1 + \|y\|^2\right\}\right) \frac{d\eta dy ds}{(1+|\log s|)^2}$$

$$= \frac{\Gamma(p/2+n/2+1)}{2^{-p/2-n/2-1}} \int_0^\infty \frac{ds}{s(1+|\log s|)^2} \int_{\mathbb{R}^p} \frac{\|y\|^2 dy}{(1+\|y\|^2)^{p/2+n/2+1}} (g+1)$$

$$= 2q_3(0)(g+1),$$

where the second equality follows from change of variables

$$y_i = \frac{x_i}{\sqrt{1+g\sqrt{s}}} \quad \text{with Jacobian} \quad |\partial x/\partial y| = (1+g)^{p/2} s^{p/2}$$

and the last equality follows from Part 3 of Lemma A.2.   □

A.7.2 Proof for A_2

For A_2, we have only to consider the case $\max(-p/2 + 1, 0) < a \leq 1$. Let $\epsilon = a/2$. Recall

$$A_2(w, s; i) = \frac{\left(\iint (g + 1)^{-1} F h_i^2 \pi (1 - k_i^2) dg d\eta \right)^2}{\iint F h_i^2 \pi dg d\eta}.$$

By the Cauchy-Schwarz inequality (Part 1 of Lemma A.3),

$$A_2(w, s; i) \leq \iint \frac{(1 - k_i^2)^2}{(g + 1)^2} F h_i^2 \pi dg d\eta. \tag{A.36}$$

The two inequalities

$$1 - k_i^2 = (1 + k_i)(1 - k_i) \leq 2(1 - k_i) = \frac{2 \log(g + 1)}{\log(g + 1 + i)},$$

$$\frac{\epsilon}{2} \log(g + 1) \leq (g + 1)^{\epsilon/2} - 1 \leq (g + 1)^{\epsilon/2},$$

give

$$1 - k_i^2 \leq \frac{2 \log(g + 1)}{\log(1 + i)} \leq \frac{4(1 + g)^{\epsilon/2}}{\epsilon \log(1 + i)}. \tag{A.37}$$

By (A.36) and (A.37),

$$A_2(w, s; i) \leq \frac{16}{\epsilon^2 \{\log(1 + i)\}^2} \iint \frac{F h_i^2 \pi dg d\eta}{(g + 1)^{2 - \epsilon}}.$$

Then, by Lemma A.12 below as well as Part 2 of Lemma 3.1, the integral involving A_2 is bounded as follows:

$$\iint \|x\|^2 s^{n/2 - 1} A_2(\|x\|^2 / s, s; i) dx ds \leq \frac{16 q_3(0) \int_0^\infty \eta^{-1} h_i^2(\eta) d\eta}{\epsilon^2 \{\log(1 + i)\}^2} \int_0^\infty \frac{\pi(g) dg}{(g + 1)^{1 - \epsilon}}$$

$$= \frac{32 q_3(0) B(a/2, b + 1)}{\epsilon^2 \log(i + 1)} < \infty,$$

where $q_3(0)$ is given by (A.35). Hence, by dominated convergence,

$$\lim_{i \to \infty} \iint \|x\|^2 s^{n/2 - 1} A_2(\|x\|^2 / s, s; i) dx ds = 0,$$

which completes the proof for A_2.

Lemma A.12 *Assume* $-n/2 < \beta < p/2 + 1$. *Then*

$$\iint \frac{\|x\|^2 s^{n/2-1}}{(\|x\|^2/s)^\beta} F(g, \eta; \|x\|^2/s, s) dx ds = \frac{q_3(\beta)}{\eta(g+1)^{\beta-1}}.$$

Proof

$$\iint \frac{\|x\|^2 s^{n/2-1}}{(\|x\|^2/s)^\beta} F(g, \eta; \|x\|^2/s, s) dx ds$$

$$= \iint \frac{\|x\|^2 s^{n/2-1}}{(\|x\|^2/s)^\beta} \frac{\eta^{p/2+n/2}}{(g+1)^{p/2}} \exp\left(-\frac{s\eta}{2}\left(\frac{\|x\|^2/s}{1+g} + 1\right)\right) dx ds$$

$$= \iint (g+1)^{p/2} s^{p/2} \frac{\|y\|^2 s(g+1) s^{n/2-1}}{\|y\|^{2\beta}(1+g)^\beta} \frac{\eta^{p/2+n/2}}{(g+1)^{p/2}} \exp\left(-\frac{s\eta}{2}\left(\|y\|^2 + 1\right)\right) dy ds$$

$$= \frac{\Gamma(p/2+n/2+1)}{2^{-p/2-n/2-1}} \int_{\mathbb{R}^p} \frac{\|y\|^{2(1-\beta)} dy}{(1+\|y\|^2)^{p/2+n/2+1}} \frac{1}{\eta(g+1)^{\beta-1}} = \frac{q_3(\beta)}{\eta(g+1)^{\beta-1}}. \quad \square$$

A.7.3 Proof for A_3

For A_3, we have only to consider the case $\max(-p/2 + 1, 0) < a \leq 1$. Let $\epsilon = a/2$. Recall

$$A_3(w, s; i) = \frac{(\iint (g+1)^{-1} F h_i^2 \pi k_i^2 dg d\eta)^2}{(\iint F h_i^2 \pi dg d\eta)^2 \iint F h_i^2 \pi k_i^2 dg d\eta} \left(\iint F h_i^2 \pi (1 - k_i^2) dg d\eta\right)^2.$$

By the Cauchy–Schwarz inequality (Part 1 of Lemma A.3),

$$\left(\iint \frac{F h_i^2 \pi k_i^2}{g+1} dg d\eta\right)^2 \leq \iint F h_i^2 \pi k_i^2 dg d\eta \iint \frac{F h_i^2 \pi k_i^2}{(g+1)^2} dg d\eta$$

$$\leq \iint F h_i^2 \pi k_i^2 dg d\eta \iint \frac{F h_i^2 \pi}{(g+1)^2} dg d\eta,$$

where the second inequality follows from $k_i^2 \leq 1$. Further

$$\frac{(\iint F h_i^2 \pi (1 - k_i^2) dg d\eta)^2}{\iint F h_i^2 \pi dg d\eta} \leq \iint (1 - k_i^2)^2 F h_i^2 \pi dg d\eta$$

$$\leq \frac{16}{\epsilon^2 \{\log(1+i)\}^2} \iint (g+1)^\epsilon F h_i^2 \pi dg d\eta,$$

where the second inequality follows from (A.37). The following lemma is useful in completing the proof.

Lemma A.13 *There exists a positive constant q_4 such that*

$$\frac{\iint (g+1)^\epsilon F h_i^2 \pi \, dg \, d\eta}{\iint F h_i^2 \pi \, dg \, d\eta} \le q_4(w^\epsilon + 1).$$

Proof See Sect. A.9. □

By Lemma A.13, we have

$$A_3(w, s; i) \le \frac{16 q_4(w^\epsilon + 1)}{\epsilon^2 \{\log(1+i)\}^2} \iint \frac{F h_i^2 \pi k_i^2}{(g+1)^2} dg \, d\eta.$$

Then, by Lemma A.12 as well as Part 2 of Lemma 3.1,

$$\iint \|x\|^2 s^{n/2-1} A_3(\|x\|^2/s, w; i) \, dx \, ds$$

$$\le \frac{32 q_4}{\epsilon^2 \log(i+1)} \left(q_3(-\epsilon) \int_0^\infty \frac{\pi(g) \, dg}{(g+1)^{1-\epsilon}} + q_3(0) \int_0^\infty \frac{\pi(g) \, dg}{g+1} \right)$$

$$\le \frac{32 q_4 \{q_3(-\epsilon) + q_3(0)\}}{\epsilon^2 \log(i+1)} B(a/2, b+1) < \infty.$$

Hence, again by dominated convergence,

$$\lim_{i \to \infty} \iint \|x\|^2 s^{n/2-1} A_3(\|x\|^2/s, s; i) \, dx \, ds = 0,$$

which completes the proof for A_3.

A.8 Proof of Lemma A.10

The proof of Lemma A.10 in Sect. A.7.1 is based on Lemmas A.14–A.16, whose proofs are given in Sects. A.8.1–A.8.4.

First we re-express $\tilde{A}_1(w, s; i)$ as follows.

Lemma A.14 *Let $z = \|x\|^2/(\|x\|^2 + s)$. Then*

$$\tilde{A}_1(\|x\|^2/s, s; i) = 1 - \frac{\{E[H_i(V/s) \mid z]\}^2}{E[H_i^2(V/s) \mid z]}, \tag{A.38}$$

where the expected value is with respect to the probability density on $v \in (0, \infty)$,

$$f(v \mid z) = \frac{v^{(p+n)/2}}{\psi(z)} \int_0^1 \frac{t^{p/2-2+a}(1-t)^b}{(1-zt)^{p/2+a+b}} \exp\left(-\frac{v}{2(1-zt)}\right) dt, \tag{A.39}$$

with normalizing constant $\psi(z)$ given by

$$\psi(z) = \iint \frac{t^{p/2-2+a}(1-t)^b}{(1-zt)^{p/2+a+b}} v^{(p+n)/2} \exp\left(-\frac{v}{2(1-zt)}\right) dv dt, \qquad (A.40)$$

and

$$H_i(\eta) = \frac{h_i(\eta)}{\log(i+1)} = \frac{1}{\log(i+1)+|\log\eta|}. \qquad (A.41)$$

The behavior of the probability density f given in (A.39) is summarized in the following lemma.

Lemma A.15 *Suppose* $-p/2+1 < a < n/2+2$.

1. *For $s \leq 1$ and for $k \geq 0$, there exist $C_1(k) > 0$ and $C_2(k) > 0$ such that*

$$s^{-C_1(k)} \int_0^s |\log v|^k f(v\,|\,z) dv \leq C_2(k).$$

2. *For $s > 1$ and for $k \geq 0$, there exists $C_3(k) > 0$ such that*

$$\exp(s/4) \int_s^\infty |\log v|^k f(v\,|\,z) dv \leq C_3(k).$$

It follows from Lemma A.15 that

$$E\left[|\log V|^k\,|\,z\right] < C_2(k) + C_3(k) := C_4(k). \qquad (A.42)$$

Using Lemma A.15, $\{E[H_i(V/s)\,|\,z]\}^2$ and $E[H_i^2(V/s)\,|\,z]$ with $H_i(\cdot)$ given in (A.41) are bounded as follows.

Lemma A.16 *Let $j = \log(i+1)$.*

1. *There exist $0 < C_5 < 1$ and $C_6 > 0$ such that*

$$(j-\log s)^2\{E[H_i(V/s)\,|\,z]\}^2 \geq 1 - 2\frac{E[\log V\,|\,z]}{j-\log s} - \frac{C_6}{(1-\log s)^2}$$

for all $0 < s < C_5$, all $z \in (0, 1)$ and all $j \geq 1$.
2. *There exists $C_7 > 0$ such that*

$$(j-\log s)^2\,E[H_i^2(V/s)\,|\,z] \leq 1 - 2\frac{E\left[\log V\,|\,z\right]}{j-\log s} + \frac{C_7}{(1-\log s)^2}$$

for all $0 < s < 1$, all $z \in (0, 1)$ and all $j \geq 1$.
3. *There exist $C_8 > 1$ and $C_9 > 0$ such that*

$$(j + \log s)^2 \{E[H_i(V/s) \mid z]\}^2 \geq 1 + 2\frac{E[\log V \mid z]}{j + \log s} - \frac{C_9}{(1 + \log s)^2}$$

for all $s > C_8$, all $z \in (0, 1)$ and all $j \geq 1$.

4. There exists $C_{10} > 0$ such that

$$(j + \log s)^2 E[H_i^2(V/s) \mid z] \leq 1 + 2\frac{E[\log V \mid z]}{j + \log s} + \frac{C_{10}}{(1 + \log s)^2}$$

for all $s > 1$, all $z \in (0, 1)$ and all $j \geq 1$.

Using Lemmas A.14–A.16, we now complete the proof of Lemma A.10. As in Lemma A.16, we still assume $j = \log(i + 1)$.

[**Proof for smaller** s] We first bound $\tilde{A}_1(w, s; i)$ for $0 < s < \gamma_1$ where γ_1 is defined by

$$\gamma_1 = \min[C_5, 1/\exp\{4C_4(1)\}].$$

Note, for $0 < s < \gamma_1$,

$$1 - 2\frac{E[\log V \mid z]}{j - \log s} \geq 1 - 2\frac{E[|\log V| \mid z]}{-\log s} \geq 1 - \frac{2C_4(1)}{4C_4(1)} = \frac{1}{2}, \quad (A.43)$$

where the second inequality follows from (A.42). Further, by Parts 1 and 2 of Lemma A.16, for $0 < s < \gamma_1$, we have

$$\tilde{A}_1(\|x\|^2/s, s; i) = 1 - \frac{\{E[H_i(V/s) \mid z]\}^2}{E[H_i^2(V/s) \mid z]} \leq 1 - \frac{1 - 2\dfrac{E[\log V \mid z]}{j - \log s} - \dfrac{C_6}{(1 - \log s)^2}}{1 - 2\dfrac{E[\log V \mid z]}{j - \log s} + \dfrac{C_7}{(1 - \log s)^2}}$$

$$= \frac{C_6 + C_7}{(1 - \log s)^2}\left(1 - 2\frac{E[\log V \mid z]}{j - \log s} + \frac{C_7}{(1 - \log s)^2}\right)^{-1} \leq \frac{2C_6 + 2C_7}{(1 - \log s)^2}$$

where the second inequality follows from (A.43).

[**Proof for larger** s] Here we bound $\tilde{A}_1(w, s; i)$ for $s > \gamma_2 > 1$ where γ_2 is defined by

$$\gamma_2 = \max\{C_8, \exp(4C_4(1))\}.$$

Note, for $s > \gamma_2$,

$$1 + 2\frac{E[\log V \mid z]}{j + \log s} \geq 1 - 2\frac{E[|\log V| \mid z]}{\log s} \geq 1 - \frac{2C_4(1)}{4C_4(1)} = \frac{1}{2}, \quad (A.44)$$

where the second inequality follows from (A.42). Further, by Parts 3 and 4 of Lemma A.16, for $s > \gamma_2$, we have

$$\tilde{A}_1(\|x\|^2/s, s; i) = 1 - \frac{\{E[H_i(V/s)\,|\,z]\}^2}{E[H_i^2(V/s)\,|\,z]} \leq 1 - \frac{1 + 2\dfrac{E\left[\log V\,|\,z\right]}{j + \log s} - \dfrac{C_9}{(1 + \log s)^2}}{1 + 2\dfrac{E\left[\log V\,|\,z\right]}{j + \log s} + \dfrac{C_{10}}{(1 + \log s)^2}}$$

$$= \frac{C_9 + C_{10}}{(1 + \log s)^2}\left(1 + 2\frac{E\left[\log V\,|\,z\right]}{j + \log s} + \frac{C_{10}}{(1 + \log s)^2}\right)^{-1} \leq \frac{2C_9 + 2C_{10}}{(1 + \log s)^2}$$

where the second inequality follows from (A.44).

By (A.38), $\tilde{A}_1(w, s; i) \leq 1$ for all x and s and thus the bound for $\gamma_1 \leq s \leq \gamma_2$ is 1. With

$$C_{11} = 2\max(C_6 + C_7, C_9 + C_{10}, 1/2)\{1 + \log\max(1/\gamma_1, \gamma_2)\}^2,$$

we have $\tilde{A}_1(w, s; i) \leq C_{11}/(1 + |\log s|)^2$ for all $s > 0$ and this completes the proof of Lemma A.10.

A.8.1   Proof of Lemma A.14

As in (A.41), $H_i(\eta) = h_i(\eta)/j$ with $j = \log(i + 1)$ and hence

$$j^{-\ell}\iint F h_i^\ell \pi\,dg\,d\eta$$

$$= \iint \frac{\eta^{p/2+n/2}}{(g + 1)^{p/2}}\exp\left\{-\frac{\eta s}{2}\left(\frac{w}{g + 1} + 1\right)\right\}H_i^\ell(\eta)(g + 1)^{-a}\left(\frac{g}{g + 1}\right)^b dg\,d\eta,$$

for $\ell = 0, 1, 2$. Apply the change of variables

$$g = \frac{1 - t}{(1 - z)t} \quad \text{where } z = \frac{w}{w + 1}$$

with

$$g + 1 = \frac{1 - zt}{(1 - z)t}, \quad 1 + \frac{w}{g + 1} = 1 + \frac{z/(1 - z)}{g + 1} = \frac{1}{1 - zt}, \quad \left|\frac{dg}{dt}\right| = \frac{1}{(1 - z)t^2}.$$

Then

$$\{\log(i+1)\}^{-\ell} \iint F h_i^\ell \pi \, dg \, d\eta \tag{A.45}$$

$$= \iint \left(\frac{(1-z)t}{1-zt}\right)^{p/2+a} \left(\frac{1-t}{1-zt}\right)^b \frac{\eta^{p/2+n/2}}{(1-z)t^2} \exp\left(-\frac{\eta s}{2(1-zt)}\right) H_i^\ell(\eta) \, dt \, d\eta$$

$$= (1-z)^{p/2-1+a} \iint \frac{t^{p/2-2+a}(1-t)^b}{(1-zt)^{p/2+a+b}} \eta^{p/2+n/2} \exp\left(-\frac{\eta s}{2(1-zt)}\right) H_i^\ell(\eta) \, dt \, d\eta$$

$$= \frac{(1-z)^{p/2-1+a}}{s^{p/2+n/2+1}} \iint \frac{t^{p/2-2+a}(1-t)^b}{(1-zt)^{p/2+a+b}} v^{p/2+n/2} \exp\left(-\frac{v}{2(1-zt)}\right) H_i^\ell(v/s) \, dt \, dv$$

$$= \frac{(1-z)^{p/2-1+a}}{s^{p/2+n/2+1}} \psi(z) \, \mathrm{E}[H_i^\ell(V/s) \mid z],$$

where $\psi(z)$ is given by (A.40), and the result follows.

A.8.2 Properties of $\psi(z)$ and $f(v \mid z)$

This section presents preliminary results for Lemma A.15. We consider a function more general than $\psi(z)$ given by (A.40). Let

$$\psi(z; \ell, m) = \iint \frac{t^{p/2-2+a}(1-t)^b}{(1-zt)^{p/2+a+b}} v^{(p+n)/2+(n/2+2-a)\ell} \exp\left(-\frac{v}{2m(1-zt)}\right) dv \, dt,$$

under the conditions

$$n/2 + 2 - a > 0, \quad \ell > -1, \quad m > 0. \tag{A.46}$$

Clearly $\psi(z)$ given by (A.40) is

$$\psi(z) = \psi(z; 0, 1).$$

Lemma A.17 *Assume assumption* (A.46). *Then*

$$0 < \psi(0; \ell, m) < \infty \quad \text{and} \quad 0 < \psi(1; \ell, m) < \infty. \tag{A.47}$$

Further

$$T_1(\ell, m) \le \psi(z; \ell, m) \le T_2(\ell, m), \tag{A.48}$$

where

$$T_1(\ell, m) = \min\{\psi(0; \ell, m), \psi(1; \ell, m)\} \text{ and}$$
$$T_2(\ell, m) = \max\{\psi(0; \ell, m), \psi(1; \ell, m)\}.$$

Proof Note

$$\psi(z; \ell, m) = \Gamma(\{p+n\}/2 + 1 + \{n/2+2-a\}\ell)(2m)^{(p+n)/2+1+(n/2+2-a)\ell}$$
$$\times \int_0^1 t^{p/2-2+a}(1-t)^b(1-zt)^{(n/2+2-a)(\ell+1)-b-1}dt,$$

which is monotone in z (either increasing or decreasing depending on the sign of $(n/2+2-a)(\ell+1)-b-1$). Further,

$$\psi(0; \ell, m) = \frac{\Gamma(\{p+n\}/2 + 1 + \{n/2+2-a\}\ell)}{(2m)^{-(p+n)/2-1-(n/2+2-a)\ell}} B(p/2 - 1 + a, b + 1),$$

$$\psi(1; \ell, m) = \frac{\Gamma(\{p+n\}/2 + 1 + \{n/2+2-a\}\ell)}{(2m)^{-(p+n)/2-1-(n/2+2-a)\ell}} B(p/2 - 1 + a, (n/2+2-a)(\ell+1)),$$

which are positive and finite under the assumption (A.46). Thus (A.47) and (A.48) follow. □

Lemma A.18 *For any $\epsilon \in (0, 1)$,*

$$f(v \mid z) \le T_3(\epsilon) v^{n/2+1-a-\epsilon(b+1)}$$

where

$$T_3(\epsilon) = \frac{\{p - 2 + 2a + 2\epsilon(b+1)\}^{p/2-1+a+\epsilon(b+1)} B(p/2 - 1 + a, (b+1)\epsilon)}{T_1(0, 1)}.$$

Proof Note, for $\epsilon \in (0, 1)$,

$$\frac{(1-t)^b}{(1-zt)^{p/2+a+b}} = (1-t)^{(b+1)\epsilon-1}\left(\frac{1-t}{1-zt}\right)^{(b+1)(1-\epsilon)}(1-zt)^{-p/2+1-a-\epsilon(b+1)}$$
$$\le (1-t)^{(b+1)\epsilon-1}(1-zt)^{-p/2+1-a-\epsilon(b+1)}.$$

Part 7 of Lemma A.5, gives

$$\frac{\exp(-v/\{2(1-zt)\})}{(1-zt)^{p/2-1+a+\epsilon(b+1)}} \le \left(\frac{p - 2 + 2a + 2\epsilon(b+1)}{v}\right)^{p/2-1+a+\epsilon(b+1)}.$$

Further, by Lemma A.17, $\psi(z) \ge T_1(0, 1)$ for all $z \in (0, 1)$. Hence, by the definition of $f(v \mid z)$ given by (A.39),

$$f(v \mid z) \le T_3(\epsilon) v^{n/2+1-a-\epsilon(b+1)}.$$ □

A.8.3 Proof of Lemma A.15

[Part 1] Let

$$\epsilon_* = \frac{1}{4}\min\left(\frac{n/2+2-a}{b+1}, 2\right) \in (0,1)$$

in Lemma A.18. Then,

$$C_1(0) := n/2 + 2 - a - \epsilon_*(b+1) = \frac{n/2+2-a}{4}\left\{4 - \min\left(1, 2\frac{b+1}{n/2+2-a}\right)\right\} > 0$$

and hence

$$\int_0^s f(v\mid z)\mathrm{d}v \le \frac{T_3(\epsilon_*)}{C_1(0)}s^{C_1(0)} = C_2(0)s^{C_1(0)}, \tag{A.49}$$

where $C_2(0)$ is defined by $C_2(0) = T_3(\epsilon_*)/C_1(0)$. Hence the lemma holds for $k = 0$. Now consider $k > 0$. By Part 4 of Lemma A.5,

$$|\log v|^k \le \left(\frac{4k}{n/2-a}\right)^k v^{-(n/2+2-a)/4},$$

for all $v \in (0,1)$. Thus, for $k > 0$,

$$C_1(k) := n/2 + 2 - a - \epsilon_*(b+1) - \frac{1}{4}(n/2+2-a)$$

$$= (n/2+2-a)\left\{\frac{3}{4} - \frac{1}{4}\min\left(1, 2\frac{b+1}{n/2+2-a}\right)\right\} > 0.$$

Hence, for $k > 0$,

$$\int_0^s |\log v|^k f(v\mid z)\mathrm{d}v \le C_2(k)s^{C_1(k)}, \tag{A.50}$$

where $C_2(k)$ is defined by

$$C_2(k) = \frac{T_3(\epsilon_*)}{C_1(k)}\left(\frac{4k}{n/2+2-a}\right)^k.$$

By (A.49) and (A.50), Part 1 follows.

[Part 2] Note, for $v \ge s$,

$$\exp\left(-\frac{v}{2(1-zt)}\right) = \exp\left(-\frac{v}{4(1-zt)}\right)\exp\left(-\frac{v}{4(1-zt)}\right)$$

$$\le \exp\left(-\frac{v}{4(1-zt)}\right)\exp\left(-\frac{v}{4}\right) \le \exp\left(-\frac{v}{4(1-zt)}\right)\exp\left(-\frac{s}{4}\right).$$

For $k = 0$, by Lemma A.17,

$$
\exp(s/4) \int_s^\infty f(v \mid z) dv
$$

$$
\leq \int_s^\infty \frac{v^{(p+n)/2}}{\psi(z; 0, 1)} \int_0^1 \frac{t^{p/2-2+a}(1-t)^b}{(1-zt)^{p/2+a+b}} \exp\left(-\frac{v}{4(1-zt)}\right) dt dv
$$

$$
\leq \frac{\psi(z, 0, 2)}{\psi(z; 0, 1)} \leq \frac{T_2(0, 2)}{T_1(0, 1)} := C_3(0),
$$

where the third inequality follows from Lemma A.17.

For $k > 0$, note by Part 5 of Lemma A.5,

$$
|\log v|^k \leq \left(\frac{2k}{n/2+2-a}\right)^k \left(v^{(n/2+2-a)/2} + v^{-(n/2+2-a)/2}\right).
$$

Then

$$
\exp(s/4) \int_s^\infty |\log v|^k f(v \mid z) dv
$$

$$
\leq \int_s^\infty \left(\frac{2k}{n/2+2-a}\right)^k \left(v^{(n/2+2-a)/2} + v^{-(n/2+2-a)/2}\right)
$$

$$
\times \frac{v^{(p+n)/2}}{\psi(z; 0, 1)} \int_0^1 \frac{t^{p/2-2+a}(1-t)^b}{(1-zt)^{p/2+a+b}} \exp\left(-\frac{v}{4(1-zt)}\right) dt dv
$$

$$
\leq \left(\frac{2k}{n/2+2-a}\right)^k \frac{\psi(z, -1/2, 2) + \psi(z, 1/2, 2)}{\psi(z; 0, 1)}
$$

$$
\leq \left(\frac{2k}{n/2+2-a}\right)^k \frac{T_2(-1/2, 2) + T_2(1/2, 2)}{T_1(0, 1)}
$$

$$
:= C_3(k),
$$

where the third inequality follows from Lemma A.17. This completes the proof. □

A.8.4 Proof of Lemma A.16

[Part 1] Assume $s < 1$ equivalently $-\log s > 0$. Then by Part 1 of Lemma A.19 below, we have

$$(j - \log s)\, \mathrm{E}\left[H_i(V/s)\,|\,z\right]$$

$$= \int_0^s \frac{j - \log s}{j + \log(s/v)} f(v\,|\,z)\mathrm{d}v + \int_s^\infty \frac{j - \log s}{j + \log(v/s)} f(v\,|\,z)\mathrm{d}v$$

$$\geq \int_s^\infty \frac{j - \log s}{j + \log(v/s)} f(v\,|\,z)\mathrm{d}v$$

$$\geq \int_s^\infty \left(1 - \frac{\log v}{j - \log s} - \frac{|\log v|^3}{(1 - \log s)^2}\right) f(v\,|\,z)\mathrm{d}v$$

$$\geq 1 - \int_0^s f(v\,|\,z)\mathrm{d}v - \frac{\mathrm{E}\left[\log V\,|\,z\right]}{j - \log s} - \frac{\int_0^s |\log v| f(v\,|\,z)\mathrm{d}v}{j - \log s} - \frac{\mathrm{E}\left[|\log V|^3\,|\,z\right]}{(1 - \log s)^2}.$$

By Lemma A.15, there exists $T_4 > 0$ such that

$$\int_0^s f(v\,|\,z)\mathrm{d}v + \frac{\int_0^s |\log v| f(v\,|\,z)\mathrm{d}v}{j - \log s} + \frac{\mathrm{E}\left[|\log V|^3\,|\,z\right]}{(1 - \log s)^2} \leq \frac{T_4}{(1 - \log s)^2}$$

for all $s \in (0, 1)$ and hence $(j - \log s)\,\mathrm{E}\left[H_i(V/s)\,|\,z\right] \geq g(s, z; i)$ where

$$g(s, z; i) = 1 - \frac{\mathrm{E}\left[\log V\,|\,z\right]}{j - \log s} - \frac{T_4}{(1 - \log s)^2}. \tag{A.51}$$

Further

$$\frac{|\mathrm{E}\left[\log V\,|\,z\right]|}{j - \log s} < \frac{\mathrm{E}\left[|\log V|\,|\,z\right]}{1 - \log s} \leq \frac{C_4(1)}{1 - \log s} \tag{A.52}$$

and hence $g(s, z; i) \geq 0$, for all $s < C_5 = 1/\exp\{C_4(1) + T_4\}$. Consider $\{g(s, z; i)\}^2$ for all $s < C_5$. By (A.51) and (A.52),

$$\{g(s, z; i)\}^2 \geq 1 - 2\frac{\mathrm{E}\left[\log V\,|\,z\right]}{j - \log s} - \frac{2T_4|\mathrm{E}\left[\log V\,|\,z\right]|}{(1 - \log s)^3} - \frac{2T_4}{(1 - \log s)^2}$$

$$\geq 1 - 2\frac{\mathrm{E}\left[\log V\,|\,z\right]}{j - \log s} - \frac{C_6}{(1 - \log s)^2},$$

where $C_6 = 2T_4\{C_4(1) + 1\}$. This completes the proof for Part 1.

[Part 2] Assume $s < 1$ equivalently $-\log s > 0$. We consider $\mathrm{E}\left[H_i^2(V/s)\,|\,z\right]$ given by

$$\mathrm{E}\left[H_i^2(V/s)\,|\,z\right] = \int_0^s \frac{f(v\,|\,z)}{\{j + \log(s/v)\}^2}\mathrm{d}v + \int_s^\infty \frac{f(v\,|\,z)}{\{j + \log(v/s)\}^2}\mathrm{d}v. \tag{A.53}$$

Note

$$(j - \log s)^2 \int_0^s \frac{f(v \mid z)}{\{j + \log(s/v)\}^2} dv \leq \frac{(j - \log s)^2}{j^2} \int_0^s f(v \mid z) dv$$

$$\leq (1 - \log s)^2 \int_0^s f(v \mid z) dv = \frac{(1 - \log s)^4 \int_0^s f(v \mid z) dv}{(1 - \log s)^2}.$$

In the numerator above, by Lemma A.15, there exists $T_5 > 0$ such that

$$(1 - \log s)^4 \int_0^s f(v \mid z) dv \leq T_5$$

for all $s \in (0, 1)$. Further, by Part 1 of Lemma A.19 below,

$$(j - \log s)^2 \int_s^\infty \frac{f(v \mid z) dv}{\{j + \log(v/s)\}^2} \leq \int_s^\infty \left(1 - \frac{2 \log v}{j - \log s} + 4 \frac{\sum_{\ell=2}^6 |\log v|^\ell}{(1 - \log s)^2} \right) f(v \mid z) dv$$

$$\leq \int_0^\infty \left(1 - \frac{2 \log v}{j - \log s} + 4 \frac{\sum_{\ell=2}^6 |\log v|^\ell}{(1 - \log s)^2} \right) f(v \mid z) dv$$

$$= 1 - 2 \frac{E[\log V \mid z]}{j - \log s} + 4 \frac{\sum_{\ell=2}^6 C_4(\ell)}{(1 - \log s)^2}. \tag{A.54}$$

Then by (A.53) and (A.54),

$$(j - \log s)^2 \, E\left[H_i^2(V/s) \mid z \right] \leq 1 - 2 \frac{E[\log V \mid z]}{j - \log s} + \frac{C_7}{(1 - \log s)^2},$$

for all $s \in (0, 1)$, where $C_7 = T_5 + 4 \sum_{\ell=2}^6 C_4(\ell)$. This completes the proof for Part 2.

[Part 3] Assume $s > 1$ equivalently $\log s > 0$. Then by Part 2 of Lemma A.19 below,

$$(j + \log s) \, E\left[H_i(V/s) \mid z \right]$$

$$= \int_0^s \frac{j + \log s}{j + \log(s/v)} f(v \mid z) dv + \int_s^\infty \frac{j + \log s}{j + \log(v/s)} f(v \mid z) dv$$

$$\geq \int_0^s \frac{j + \log s}{j + \log(s/v)} f(v \mid z) dv$$

$$\geq \int_0^s \left(1 + \frac{\log v}{j + \log s} - \frac{|\log v|^3}{(1 + \log s)^2} \right) f(v \mid z) dv$$

$$\geq 1 - \int_s^\infty f(v \mid z) dv + \frac{E\left[\log V \mid z \right]}{j + \log s} - \frac{\int_s^\infty |\log v| f(v \mid z) dv}{j + \log s} - \frac{E\left[|\log V|^3 \mid z \right]}{(1 + \log s)^2}.$$

By Lemma A.15, there exists $T_6 > 0$ such that

$$\int_s^\infty f(v\mid z)dv + \frac{\int_s^\infty |\log v| f(v\mid z)dv}{j+\log s} + \frac{\mathrm{E}\left[|\log V|^3\mid z\right]}{(j+\log s)^2} \le \frac{T_6}{(1+\log s)^2},$$

for all $s \in (1, \infty)$ and hence

$$(j+\log s)\,\mathrm{E}\left[H_i(V/s)\mid z\right] \ge g(s, z; i)$$

where

$$g(s, z; i) = 1 + \frac{\mathrm{E}\left[\log V\mid z\right]}{j+\log s} - \frac{T_6}{(1+\log s)^2}.$$

Further, by (A.52), we have $g(s, z; i) \ge 0$, for all $s > C_8$ where $C_8 = \exp\{C_4(1) + T_6\}$. Now consider $\{g(s, z; i)\}^2$ for all $s > C_8$. By (A.51) and (A.52),

$$\{g(s, z; i)\}^2 \ge 1 + 2\frac{\mathrm{E}\left[\log V\mid z\right]}{j+\log s} - \frac{2T_6\,\mathrm{E}\left[\log V\mid z\right]}{(1+\log s)^3} - \frac{2T_6}{(1+\log s)^2}$$

$$\ge 1 + 2\frac{\mathrm{E}\left[\log V\mid z\right]}{j+\log s} - \frac{C_9}{(1+\log s)^2},$$

where $C_9 = 2T_6\{C_4(1) + 1\}$. This completes the proof for Part 3.

[Part 4] Assume $s > 1$ equivalently $\log s > 0$. We consider $\mathrm{E}\left[H_i^2(V/s)\mid z\right]$ given by

$$\mathrm{E}\left[H_i^2(V/s)\mid z\right] = \int_0^s \frac{f(v\mid z)}{\{j+\log(s/v)\}^2}dv + \int_s^\infty \frac{f(v\mid z)}{\{j+\log(v/s)\}^2}dv. \quad (A.55)$$

Note

$$(j+\log s)^2 \int_s^\infty \frac{f(v\mid z)}{\{j+\log(v/s)\}^2}dv \le \frac{(j+\log s)^2}{j^2} \int_s^\infty f(v\mid z)dv$$

$$\le (1+\log s)^2 \int_s^\infty f(v\mid z)dv = \frac{(1+\log s)^4 \int_s^\infty f(v\mid z)dv}{(1+\log s)^2}.$$

In the numerator above, by Lemma A.15, there exists $T_7 > 0$ such that

$$(1+\log s)^4 \int_s^\infty f(v\mid z)dv \le T_7$$

for all $s \in (1, \infty)$. Further, by Part 2 of Lemma A.19 below,

$$(j + \log s)^2 \int_0^s \frac{f(v \mid z)}{\{j + \log(s/v)\}^2} dv$$

$$\leq \int_0^s \left(1 + \frac{2 \log v}{j + \log s} + 4 \frac{\sum_{\ell=2}^6 |\log v|^\ell}{(1 + \log s)^2}\right) f(v \mid z) dv$$

$$\leq \int_0^\infty \left(1 + \frac{2 \log v}{j + \log s} + 4 \frac{\sum_{\ell=2}^6 |\log v|^\ell}{(1 + \log s)^2}\right) f(v \mid z) dv$$

$$= 1 + 2 \frac{\mathrm{E}[\log V \mid z]}{j + \log s} + 4 \frac{\sum_{\ell=2}^6 \mathrm{E}[|\log V|^\ell \mid z]}{(1 + \log s)^2}. \tag{A.56}$$

Then by (A.55) and (A.56),

$$(j + \log s)^2 \, \mathrm{E}\left[H_i^2(V/s) \mid z\right] \leq 1 + 2 \frac{\mathrm{E}[\log V \mid z]}{j + \log s} + \frac{C_{10}}{(1 + \log s)^2}$$

for all $s \in (1, \infty)$, where $C_{10} = T_7 + 4 \sum_{\ell=2}^6 C_4(\ell)$. This completes the proof for Part 4.

Lemma A.19 *Assume $j \geq 1$.*

1. *For $s < 1$ and $v \geq s$,*

$$\frac{j - \log s}{j + \log v - \log s} \geq 1 - \frac{\log v}{j - \log s} - \frac{|\log v|^3}{(1 - \log s)^2},$$

$$\frac{j - \log s}{j + \log v - \log s} \leq 1 - \frac{\log v}{j - \log s} + \frac{|\log v|^2 + |\log v|^3}{(1 - \log s)^2}$$

and

$$\left(\frac{j - \log s}{j + \log v - \log s}\right)^2 \leq 1 - 2 \frac{\log v}{j - \log s} + 4 \frac{\sum_{\ell=2}^6 |\log v|^\ell}{(1 - \log s)^2}.$$

2. *For $s > 1$ and $v \leq s$,*

$$\frac{j + \log s}{j + \log s - \log v} \geq 1 + \frac{\log v}{j + \log s} - \frac{|\log v|^3}{(1 + \log s)^2}$$

$$\frac{j + \log s}{j + \log s - \log v} \leq 1 + \frac{\log v}{j + \log s} + \frac{|\log v|^2 + |\log v|^3}{(1 + \log s)^2},$$

and

$$\left(\frac{j + \log s}{j + \log s - \log v}\right)^2 \leq 1 + 2 \frac{\log v}{j + \log s} + 4 \frac{\sum_{\ell=2}^6 |\log v|^\ell}{(1 + \log s)^2}.$$

Proof [Part 1] For $v \geq s$,

$$\frac{j - \log s}{j + \log v - \log s} = 1 - \frac{\log v}{j + \log v - \log s}$$

$$= 1 - \frac{\log v}{j - \log s} + \frac{(\log v)^2}{(j - \log s)(j + \log v - \log s)}$$

$$= 1 - \frac{\log v}{j - \log s} + \frac{(\log v)^2}{(j - \log s)^2}$$

$$- \frac{(\log v)^3}{(j - \log s)^2(j + \log v - \log s)}.$$

Then, for $j \geq 1$, three inequalities follow from the inequality

$$\left| \frac{(\log v)^3}{(j - \log s)^2(j + \log v - \log s)} \right| \leq \frac{|\log v|^3}{(j - \log s)^2} \leq \frac{|\log v|^3}{(1 - \log s)^2}.$$

[Part 2] The proof of Part 2 is similar. □

A.9 Proof of Lemma A.13

The proof of Lemma A.13 in Sect. A.7.3 is based on Lemma A.16 in Sect. A.8. Note $(g + 1)^{\epsilon} \pi(g) = (g + 1)^{-a+\epsilon} \{g/(g + 1)\}^b$. By the definition of H_i given by (A.41) and (A.45), we have

$$\frac{\iint (g + 1)^{\epsilon} F h_i^2 \pi \mathrm{d}g \mathrm{d}\eta}{\iint F h_i^2 \pi \mathrm{d}g \mathrm{d}\eta} = \frac{1}{(1 - z)^{\epsilon}} \frac{\psi(z; a - \epsilon)}{\psi(z; a)} \frac{\mathrm{E}[H_i^2(V/s) \mid z, a - \epsilon]}{\mathrm{E}[H_i^2(V/s) \mid z, a]}$$

$$= (w + 1)^{\epsilon} \frac{\psi(z; a - \epsilon)}{\psi(z; a)} \frac{\{j + |\log s|\}^2 \mathrm{E}[H_i^2(V/s) \mid z, a - \epsilon]}{\{j + |\log s|\}^2 \mathrm{E}[H_i^2(V/s) \mid z, a]}. \tag{A.57}$$

As in (A.57), in this subsection only, we write $\psi(z)$ and $\mathrm{E}[H_i^2(V/s) \mid z]$ as a function of a unlike (A.45) in Lemma A.14. By (A.48) in Lemma A.17,

$$0 < \frac{\psi(z; a - \epsilon)}{\psi(z; a)} < \infty, \tag{A.58}$$

where both (finite) lower and upper bounds are independent of z. Note

$$\frac{j + |\log s|}{j + |\log v/s|} \geq \frac{j + |\log s|}{j + |\log s| + |\log v|} \geq \frac{1}{1 + |\log v|},$$

for $j \geq 1$. By Jensen's inequality,

$$\{j + |\log s|\}^2 \, E[H_i^2(V/s) \,|\, z, a] \geq \left(\frac{1}{1 + E[|\log V \,||z]}\right)^2$$

$$\geq \left(\frac{1}{1 + \max_z E[|\log V \,||z]}\right)^2 = \left(\frac{1}{1 + C_4(1, a)}\right)^2 \qquad (A.59)$$

where $C_4(k, a)$ is given in (A.42). Further, by Parts 2 and 4 of Lemma A.16,

$$\{j + |\log s|\}^2 \, E[H_i^2(V/s) \,|\, z, a - \epsilon]$$
$$\leq 1 + 2 E[|\log V \,||z, a - \epsilon] + \max(C_7(a - \epsilon), C_{10}(a - \epsilon))$$
$$\leq 1 + 2C_4(1, a - \epsilon) + \max(C_7(a - \epsilon), C_{10}(a - \epsilon)).$$

Part 2 of Lemma A.5 gives

$$(w + 1)^\epsilon \leq w^\epsilon + 1 \quad \text{for} \quad 0 < \epsilon < 1. \qquad (A.60)$$

Combining (A.57), (A.58), (A.59) and (A.60), completes the proof of Lemma A.13.

A.10   Proof of Lemma 3.2

[Part 1] Under the assumptions on $\xi(g)$, AS.5 and AS.6, there exist $M_1 > 0$ and $w_1 > 0$ such that

$$0 \leq (g + 1)\log(g + 1)\frac{\xi'(g)}{\xi(g)} \leq M_1, \quad \text{or} \quad -M_1 \leq (g + 1)\log(g + 1)\frac{\xi'(g)}{\xi(g)} \leq 0,$$

for all $g \geq w_1$. The representation theorem for slowly varying function (Theorem 1.5 of Geluk and de Haan 1987) guarantees that there exist measurable functions $c_1(\cdot)$ and $c_2(\cdot)$ such that

$$\lim_{g \to \infty} c_1(g) = c_{1*} > 0, \quad \text{and} \quad \lim_{s \to \infty} c_2(s) = 0$$

and

$$\xi(g) = c_1(g) \exp\left(\int_1^g \frac{c_2(s)}{s} ds\right)$$

for $g \geq 1$. Recall $p/2 + a + 1 > 0$. Hence there exists an $M_2 > 0$ such that

$$\xi(g) \leq M_2(g + 1)^{(p/2+a+1)/4} \qquad (A.61)$$

for $g \geq 1$. Let

$$\epsilon = \frac{p/2 + a}{p/2 + n/2 + 1} \in (0, 1).$$

For $w \geq (w_1 + 1)^{1/(1-\epsilon)}$, let

$$I_1 = (0, 1], \quad I_2 = (1, w^{1-\epsilon} - 1],$$
$$I_3 = (w^{1-\epsilon} - 1, w^{3/2} - 1], \quad I_4 = (w^{3/2} - 1, \infty).$$

For the integral over I_1 we have

$$\int_{I_1} \frac{(g+1)^{-p/2-a}\xi(g)dg}{\{1+w/(g+1)\}^{p/2+n/2+1}} \leq \frac{\int_0^1 \xi(g)dg}{(1+w/2)^{p/2+n/2+1}}.$$

Since $p/2 + n/2 + 1 = (p/2 - 1 + a + 1/2) + (n/2 + 1 - a) + 1/2$,

$$\lim_{w\to\infty} w^{p/2-1+a+1/2} \int_{I_1} \frac{(g+1)^{-p/2-a}\xi(g)dg}{\{1+w/(g+1)\}^{p/2+n/2+1}} = 0. \tag{A.62}$$

For the integral over I_2, we have

$$\int_{I_2} \frac{(g+1)^{-p/2-a}\xi(g)dg}{\{1+w/(g+1)\}^{p/2+n/2+1}} \leq M_2 \int_{I_2} \frac{(g+1)^{-3(p/2-1+a)/4-1}dg}{(1+w^\epsilon)^{p/2+n/2+1}}$$
$$\leq \frac{4M_2}{3(p/2-1+a)} \frac{1}{w^{p/2+a}},$$

where the first inequality follows from (A.61). Hence

$$\lim_{w\to\infty} w^{p/2-1+a+1/2} \int_{I_2} \frac{(g+1)^{-p/2-a}\xi(g)dg}{\{1+w/(g+1)\}^{p/2+n/2+1}} = 0. \tag{A.63}$$

For the integral under I_4, by (A.61), we have

$$\int_{I_4} \frac{(g+1)^{-p/2-a}\xi(g)dg}{\{1+w/(g+1)\}^{p/2+n/2+1}} \leq M_2 \int_{w^{3/2}-1}^{\infty} \frac{dg}{(g+1)^{3(p/2-1+a)/4+1}}$$
$$= \frac{4M_2}{3(p/2-1+a)} \frac{1}{w^{9(p/2-1+a)/8}}.$$

Hence

$$\lim_{w\to\infty} w^{p/2-1+a+(p/2-1+a)/9} \int_{I_4} \frac{(g+1)^{-p/2-a}\xi(g)dg}{\{1+w/(g+1)\}^{p/2+n/2+1}} = 0. \tag{A.64}$$

For the integral over I_3, a change of variables gives

$$\int_{I_3} \frac{(g+1)^{-p/2-a}\xi(g)dg}{\{1+w/(g+1)\}^{p/2+n/2+1}} \tag{A.65}$$

$$= \frac{\xi(w)}{w^{p/2-1+a}} \int_{w^{-1/2}}^{w^\epsilon} \frac{t^{p/2+a-2}\xi(w/t-1)}{(1+t)^{p/2+n/2+1}\xi(w)}dt.$$

Since $\xi(g)$ is slowly varying, $\lim_{w\to\infty} \xi(w/t-1)/\xi(w) = 1$, for any fixed t. Recall $\xi(g)$ is ultimately monotone as in AS.5. Suppose $\xi(g)$ for $g \geq (w_1+1)^{1/(1-\epsilon)}$ is monotone non-decreasing. Then we

$$\int_{w^{1-\epsilon}-1}^{w^{3/2}-1} \frac{\xi'(g)}{\xi(g)}dg \leq M_1 \int_{w^{1-\epsilon}-1}^{w^{3/2}-1} \frac{dg}{(g+1)\log(g+1)}$$

which implies that

$$\frac{\xi(w^{3/2}-1)}{\xi(w^{1-\epsilon}-1)} \leq \left(\frac{3}{2(1-\epsilon)}\right)^{M_1}.$$

Similarly suppose $\xi(g)$ for $g \geq (w_1+1)^{1/(1-\epsilon)}$ is monotone non-increasing. Thus

$$\frac{\xi(w^{1-\epsilon}-1)}{\xi(w^{3/2}-1)} \leq \left(\frac{3}{2(1-\epsilon)}\right)^{M_1}.$$

Hence, for $t \in (w^{-1/2}, w^\epsilon)$,

$$\frac{\xi(w/t-1)}{\xi(w)} \leq \max\left(\frac{\xi(w^{3/2}-1)}{\xi(w^{1-\epsilon}-1)}, \frac{\xi(w^{1-\epsilon}-1)}{\xi(w^{3/2}-1)}\right) \leq \left(\frac{3}{2(1-\epsilon)}\right)^{M_1}.$$

Then, by the dominated convergence theorem,

$$\lim_{w\to\infty} \int_{w^{-1/2}}^{w^{\epsilon_1}} \frac{t^{p/2-2+a}}{(1+t)^{p/2+n/2+1}} \frac{\xi(w/t-1)}{\xi(w)}dt = \int_0^\infty \frac{t^{p/2-2+a}dt}{(1+t)^{p/2+n/2+1}}$$
$$= B(p/2-1+a, n/2+1-a),$$

and hence, by (A.65),

$$\lim_{w\to\infty} \frac{w^{p/2-1+a}}{\xi(w)} \int_{I_3} \frac{(g+1)^{-p/2-a}\xi(g)dg}{\{1+w/(g+1)\}^{p/2+n/2+1}} = B\left(\frac{p}{2}-1+a, \frac{n}{2}+2-a\right). \tag{A.66}$$

By (A.62), (A.63), (A.64) and (A.66),

$$\lim_{w\to\infty} \frac{\int_0^\infty (g+1)^{-p/2-a}\{1+w/(g+1)\}^{-(p/2+n/2+1)}\xi(g)dg}{w^{-p/2+1-a}\xi(w)B(p/2-1+a, n/2+2-a)} = 1,$$

which completes the proof of Part 1.

[Part 2] As in Part 1 of this lemma,

$$\lim_{w \to \infty} \frac{\int_0^\infty (g+1)^{-p/2-a-1}\{1 + w/(g+1)\}^{-(p/2+n/2+1)}\xi(g)dg}{w^{-p/2-a}\xi(w)B(p/2+a, n/2+1-a)} = 1.$$

Then

$$\lim_{w \to \infty} \phi(w) = \lim_{w \to \infty} w \frac{\int_0^\infty (g+1)^{-p/2-1-a}\{1 + w/(g+1)\}^{-(p/2+n/2+1)}\xi(g)dg}{\int_0^\infty (g+1)^{-p/2-a}\{1 + w/(g+1)\}^{-(p/2+n/2+1)}\xi(g)dg}$$

$$= \frac{B(p/2+a, n/2+1-a)}{B(p/2-1+a, n/2+2-a)} = \frac{p/2-1+a}{n/2+1-a},$$

which completes the proof of Part 2.

[Part 3] Generally, for differentiable functions $A(w)$ and $B(w)$, we have

$$\frac{\{wA(w)/B(w)\}'}{A(w)/B(w)} = 1 + w\frac{A'(w)}{A(w)} - w\frac{B'(w)}{B(w)}. \tag{A.67}$$

By the definition of $\phi(w)$, we have

$$\frac{w\phi'(w)}{\phi(w)} = 1 + mw\left\{-\frac{\int_0^\infty (g+1)^{-p/2-2-a}\{1 + w/(g+1)\}^{-m-1}\xi(g)dg}{\int_0^\infty (g+1)^{-p/2-1-a}\{1 + w/(g+1)\}^{-m}\xi(g)dg} \right. \tag{A.68}$$

$$\left. + \frac{\int_0^\infty (g+1)^{-p/2-1-a}\{1 + w/(g+1)\}^{-m-1}\xi(g)dg}{\int_0^\infty (g+1)^{-p/2-a}\{1 + w/(g+1)\}^{-m}\xi(g)dg}\right\},$$

where $m = p/2 + n/2 + 1$. By Part 1 of this lemma,

$$\lim_{w \to \infty} w\frac{\int_0^\infty (g+1)^{-p/2-2-a}\{1 + w/(g+1)\}^{-m-1}\xi(g)dg}{\int_0^\infty (g+1)^{-p/2-1-a}\{1 + w/(g+1)\}^{-m}\xi(g)dg}$$

$$= \frac{B(p/2+1+a, n/2+1-a)}{B(p/2+a, n/2+1-a)} = \frac{p/2+a}{p/2+n/2+1} = \frac{p/2+a}{m}$$

and

$$\lim_{w \to \infty} w\frac{\int_0^\infty (g+1)^{-p/2-1-a}\{1 + w/(g+1)\}^{-m-1}\xi(g)dg}{\int_0^\infty (g+1)^{-p/2-a}\{1 + w/(g+1)\}^{-m}\xi(g)dg}$$

$$= \frac{B(p/2+a, n/2+2-a)}{B(p/2-1+a, n/2+2-a)} = \frac{p/2-1+a}{p/2+n/2+1} = \frac{p/2-1+a}{m}.$$

Hence

$$\lim_{w \to \infty} \frac{w\phi'(w)}{\phi(w)} = 1 - (p/2+a) + (p/2-1+a) = 0,$$

which completes the proof of Part 3.

A.11   Proof of Lemma 3.3

This section is devoted to proving the equalities in Lemma 3.3.

[Proof of (3.59) of Lemma 3.3] As in (A.25) and (A.27), let

$$\varphi(g; w) = 1 - \left(\frac{g + 1 + w}{(g + 1)(w + 1)} \right)^{n/2+1}$$

$$\tilde{\varphi}(g; w) = b \frac{g + 1 + w}{g + 1} \frac{\varphi(g; w)}{g} + (p/2 - 1)\{1 - \varphi(g; w)\}.$$

As in the identity (A.28), we have

$$(n/2 + 1)w \int_0^\infty \frac{(g + 1)^{-p/2-1}\xi(g)dg}{\{1 + w/(g + 1)\}^{(p+n)/2+1}}$$

$$= (p/2 - 1) \int_0^\infty \frac{(g + 1)^{-p/2}\xi(g)dg}{\{1 + w/(g + 1)\}^{(p+n)/2+1}} - \int_0^\infty \frac{\tilde{\varphi}(g; w)(g + 1)^{-p/2}\xi(g)dg}{\{1 + w/(g + 1)\}^{(p+n)/2+1}}$$

$$+ c \int_0^\infty \varphi(g; w) \frac{1 + w/(g + 1)}{\log(g + 1) + 1} \frac{(g + 1)^{-p/2}\xi(g)dg}{\{1 + w/(g + 1)\}^{(p+n)/2+1}},$$

and hence

$$(n + 2)\phi(w) - (p - 2) \tag{A.69}$$

$$= -2 \frac{\int_0^\infty \tilde{\varphi}(g; w)(g + 1)^{-p/2}\{1 + w/(g + 1)\}^{-(p+n)/2-1}\xi(g)dg}{\int_0^\infty (g + 1)^{-p/2}\{1 + w/(g + 1)\}^{-(p+n)/2-1}\xi(g)dg}$$

$$+ 2c \frac{\int_0^\infty \frac{1 + w/(g + 1)}{\log(g + 1) + 1} \frac{\varphi(g; w)(g + 1)^{-p/2}\xi(g)dg}{\{1 + w/(g + 1)\}^{(p+n)/2+1}}}{\int_0^\infty (g + 1)^{-p/2}\{1 + w/(g + 1)\}^{-(p+n)/2-1}\xi(g)dg}.$$

Part 2 of Lemma A.9 and Part 1 of Lemma 3.2 give

$$\lim_{w \to \infty} w^{1/9} \frac{\int_0^\infty \frac{|\tilde{\varphi}(g; w)|(g + 1)^{-p/2}\xi(g)dg}{\{1 + w/(g + 1)\}^{(p+n)/2+1}}}{\int_0^\infty \frac{(g + 1)^{-p/2}\xi(g)dg}{\{1 + w/(g + 1)\}^{(p+n)/2+1}}} = 0. \tag{A.70}$$

Further Part 1 of Lemma 3.2 gives

$$\lim_{w\to\infty} \log w \frac{\displaystyle\int_0^\infty \frac{1+w/(g+1)}{\log(g+1)+1} \frac{\varphi(g;w)(g+1)^{-p/2}\xi(g)dg}{\{1+w/(g+1)\}^{(p+n)/2+1}}}{\displaystyle\int_0^\infty \frac{(g+1)^{-p/2}\xi(g)dg}{\{1+w/(g+1)\}^{(p+n)/2+1}}} \tag{A.71}$$

$$= 1 + \frac{B(p/2, n/2+1)}{B(p/2-1, n/2+2)} = 1 + \frac{p-2}{n+2}.$$

By (A.69), (A.70) and (A.71),

$$\lim_{w\to\infty} \log w \left\{ \frac{p-2}{n+2} - \phi(w) \right\} = -c\frac{2(p+n)}{(n+2)^2},$$

thus proving (3.59).

[Proof of (3.60) of Lemma 3.3] Recall (A.68) as

$$\frac{w\phi'(w)}{\phi(w)} = 1 + mw \left\{ -\frac{\int_0^\infty (g+1)^{-p/2-2}\{1+w/(g+1)\}^{-m-1}\xi(g)dg}{\int_0^\infty (g+1)^{-p/2-1}\{1+w/(g+1)\}^{-m}\xi(g)dg} \right. \tag{A.72}$$

$$\left. + \frac{\int_0^\infty (g+1)^{-p/2-1}\{1+w/(g+1)\}^{-m-1}\xi(g)dg}{\int_0^\infty (g+1)^{-p/2}\{1+w/(g+1)\}^{-m}\xi(g)dg} \right\},$$

where $m = p/2 + n/2 + 1$ and $\xi(g) = \{g/(g+1)\}^b\{\log(g+1)+1\}^{-c}$ for $b > -1$ and $c \in \mathbb{R}$. Note

$$\frac{mw}{(g+1)^2}\{1+w/(g+1)\}^{-m-1} = \frac{d}{dg}\{\{1+w/(g+1)\}^{-m} - (1+w)^{-m}\}$$

$$= \frac{d}{dg}\{\{1+w/(g+1)\}^{-m}\varphi(g;w)\},$$

where

$$\varphi(g;w) = 1 - \left(\frac{g+1+w}{(g+1)(w+1)}\right)^m,$$

and

$$\lim_{g\to 0}\{1+w/(g+1)\}^{-m}\varphi(g;w)\left(\frac{g}{g+1}\right)^b \frac{1}{\{\log(g+1)+1\}^c} = 0,$$

for $b > -1$ and $c \in \mathbb{R}$. Then an integration by parts gives

$$mw \int_0^\infty (g+1)^{-p/2-2}\{1 + w/(g+1)\}^{-m-1}\xi(g)dg$$

$$= \frac{p}{2} \int_0^\infty \frac{(g+1)^{-p/2-1}\xi(g)dg}{\{1 + w/(g+1)\}^m} - \int_0^\infty \frac{(g+1)^{-p/2-1}\tilde{\varphi}(g; w)\xi(g)dg}{\{1 + w/(g+1)\}^m}$$

$$+ c \int_0^\infty \frac{(g+1)^{-p/2-1}\varphi(g; w)\xi(g)}{\{\log(g+1) + 1\}\{1 + w/(g+1)\}^m}dg,$$

where

$$\tilde{\varphi}(g; w) = b\frac{\varphi(g; w)}{g} + \frac{p}{2}\{1 - \varphi(g; w)\}.$$

Then, by Part 1 of this lemma,

$$\lim_{w \to \infty} \log w \left(mw \frac{\int_0^\infty \dfrac{(g+1)^{-p/2-2}\xi(g)}{\{1 + w/(g+1)\}^{m+1}}dg}{\int_0^\infty \dfrac{(g+1)^{-p/2-1}\xi(g)}{\{1 + w/(g+1)\}^m}dg} - \frac{p}{2} \right) = -c. \qquad (A.73)$$

Similarly,

$$\lim_{w \to \infty} \log w \left(mw \frac{\int_0^\infty \dfrac{(g+1)^{-p/2-1}\xi(g)}{\{1 + w/(g+1)\}^{m+1}}dg}{\int_0^\infty \dfrac{(g+1)^{-p/2}\xi(g)}{\{1 + w/(g+1)\}^m}dg} - \frac{p-2}{2} \right) = -c. \qquad (A.74)$$

Combining (A.72), (A.73) and (A.74), completes the proof.

A.12 Proof of Lemma 3.4

By Part 1 of Lemma A.9, an integration by parts gives

$$(n/2 + 1 - a)w \int_0^\infty \frac{(g+1)^{-p/2-1-a}\xi(g)dg}{\{1+w/(g+1)\}^{p/2+n/2+1}}$$

$$= \left[\left(1 - \frac{w}{g+1+w}\right)^{n/2+1-a} \frac{\xi(g)}{(g+1+w)^{p/2-1+a}}\right]_0^\infty$$

$$+ (p/2 - 1 + a) \int_0^\infty \frac{(g+1)^{-p/2-a}\xi(g)dg}{\{1+w/(g+1)\}^{p/2+n/2+1}}$$

$$- \int_0^\infty (g+1)\frac{\xi'(g)}{\xi(g)}\left(1 + \frac{w}{g+1}\right)\frac{(g+1)^{-p/2-a}\xi(g)dg}{\{1+w/(g+1)\}^{p/2+n/2+1}}$$

$$= -\frac{\xi(0)}{(w+1)^{(p+n)/2}} + (p/2 - 1 + a)\int_0^\infty \frac{(g+1)^{-p/2-a}\xi(g)dg}{\{1+w/(g+1)\}^{p/2+n/2+1}}$$

$$- \int_0^\infty \Xi(g)\left(1 + \frac{w}{g+1}\right)\frac{(g+1)^{-p/2-a}\xi(g)dg}{\{1+w/(g+1)\}^{p/2+n/2+1}}.$$

By (2.29) and Lemma 2.1, we have $\Xi(g) \geq -\Xi_2(g) \geq -\Xi_{2*}$,

$$(n/2 + 1 - a)w \int_0^\infty \frac{(g+1)^{-p/2-1-a}\xi(g)dg}{\{1+w/(g+1)\}^{p/2+n/2+1}}$$

$$\leq (p/2 - 1 + a + \Xi_{2*})\int_0^\infty \frac{(g+1)^{-p/2-a}\xi(g)dg}{\{1+w/(g+1)\}^{p/2+n/2+1}}$$

$$+ \Xi_{2*}w \int_0^\infty \frac{(g+1)^{-p/2-1-a}\xi(g)dg}{\{1+w/(g+1)\}^{p/2+n/2+1}},$$

and hence

$$(n/2 + 1 - a)\phi(w) \leq (p/2 - 1 + a + \Xi_{2*}) + \Xi_{2*}\phi(w),$$

which completes the proof of the first equality.

As in (A.68),

$$\frac{w\phi'(w)}{\phi(w)} = 1 + mw\left\{-\frac{\int_0^\infty (g+1)^{-p/2-2-a}\{1+w/(g+1)\}^{-m-1}\xi(g)dg}{\int_0^\infty (g+1)^{-p/2-1-a}\{1+w/(g+1)\}^{-m}\xi(g)dg}\right. \tag{A.75}$$

$$\left. + \frac{\int_0^\infty (g+1)^{-p/2-1-a}\{1+w/(g+1)\}^{-m-1}\xi(g)dg}{\int_0^\infty (g+1)^{-p/2-a}\{1+w/(g+1)\}^{-m}\xi(g)dg}\right\},$$

where $m = p/2 + n/2 + 1$. For the second term of the right hand side of (A.75),

$$mw \int_0^\infty \frac{(g+1)^{-p/2-2-a}\xi(g)dg}{\{1+w/(g+1)\}^{m+1}} = \left[\frac{(g+1)^{-p/2-a}\xi(g)}{\{1+w/(g+1)\}^m}\right]_0^\infty \tag{A.76}$$

$$+ (p/2 + a)\int_0^\infty \frac{(g+1)^{-p/2-1-a}\xi(g)dg}{\{1+w/(g+1)\}^m} - \int_0^\infty \frac{(g+1)^{-p/2-a}\xi'(g)dg}{\{1+w/(g+1)\}^m}.$$

Similarly, for the third term of the right hand side of (A.75), we have

$$
mw \int_0^\infty \frac{(g+1)^{-p/2-1-a}\xi(g)dg}{\{1+w/(g+1)\}^{m+1}} = \left[\frac{(g+1)^{-p/2+1-a}\xi(g)}{\{1+w/(g+1)\}^m} \right]_0^\infty \tag{A.77}
$$
$$
+ (p/2-1+a) \int_0^\infty \frac{(g+1)^{-p/2-a}\xi(g)dg}{\{1+w/(g+1)\}^m} - \int_0^\infty \frac{(g+1)^{-p/2+1-a}\xi'(g)dg}{\{1+w/(g+1)\}^m}.
$$

Hence, by (A.75), (A.76) and (A.77),

$$
\frac{w\phi'(w)}{\phi(w)} = \frac{\xi(0)(1+w)^{-m}}{\int_0^\infty (g+1)^{-p/2-1-a}\{1+w/(g+1)\}^{-m}\xi(g)dg} \tag{A.78}
$$
$$
- \frac{\xi(0)(1+w)^{-m}}{\int_0^\infty (g+1)^{-p/2-a}\{1+w/(g+1)\}^{-m}\xi(g)dg}
$$
$$
+ \frac{\int_0^\infty (g+1)^{-p/2-a}\{1+w/(g+1)\}^{-m}\xi'(g)dg}{\int_0^\infty (g+1)^{-p/2-1-a}\{1+w/(g+1)\}^{-m}\xi(g)dg}
$$
$$
- \frac{\int_0^\infty (g+1)^{-p/2+1-a}\{1+w/(g+1)\}^{-m}\xi'(g)dg}{\int_0^\infty (g+1)^{-p/2-a}\{1+w/(g+1)\}^{-m}\xi(g)dg}.
$$

In the first and second terms of the right hand side of (A.78),

$$
\int_0^\infty \frac{(g+1)^{-p/2-a-1}\xi(g)dg}{\{1+w/(g+1)\}^m} \le \int_0^\infty \frac{(g+1)^{-p/2-a}\xi(g)dg}{\{1+w/(g+1)\}^m}
$$

and hence

$$
\frac{w\phi'(w)}{\phi(w)} \ge \frac{\int_0^\infty \Xi(g)(g+1)^{-p/2-1-a}\{1+w/(g+1)\}^{-m}\xi(g)dg}{\int_0^\infty (g+1)^{-p/2-1-a}\{1+w/(g+1)\}^{-m}\xi(g)dg} \tag{A.79}
$$
$$
- \frac{\int_0^\infty \Xi(g)(g+1)^{-p/2-a}\{1+w/(g+1)\}^{-m}\xi(g)dg}{\int_0^\infty (g+1)^{-p/2-a}\{1+w/(g+1)\}^{-m}\xi(g)dg}.
$$

Recall $\Xi(g) = \Xi_1(g) - \Xi_2(g)$ where $\Xi_1(g)$ is monotone non-increasing and $0 \le \Xi_2(g) \le \Xi_{2*}$, as in (2.29) and Lemma 2.1, By the correlation inequality (Lemma A.4),

$$
\frac{\int_0^\infty \Xi_1(g)(g+1)^{-p/2-1-a}\{1+w/(g+1)\}^{-m}\xi(g)dg}{\int_0^\infty (g+1)^{-p/2-1-a}\{1+w/(g+1)\}^{-m}\xi(g)dg} \tag{A.80}
$$
$$
- \frac{\int_0^\infty \Xi_1(g)(g+1)^{-p/2-a}\{1+w/(g+1)\}^{-m}\xi(g)dg}{\int_0^\infty (g+1)^{-p/2-a}\{1+w/(g+1)\}^{-m}\xi(g)dg} \ge 0.
$$

Further,

$$-\frac{\int_0^\infty \Xi_2(g)(g+1)^{-p/2-1-a}\{1+w/(g+1)\}^{-m}\xi(g)dg}{\int_0^\infty (g+1)^{-p/2-1-a}\{1+w/(g+1)\}^{-m}\xi(g)dg}$$

$$+\frac{\int_0^\infty \Xi_2(g)(g+1)^{-p/2-a}\{1+w/(g+1)\}^{-m}\xi(g)dg}{\int_0^\infty (g+1)^{-p/2-a}\{1+w/(g+1)\}^{-m}\xi(g)dg} \geq -\Xi_{2*}. \tag{A.81}$$

Combining (A.79), (A.80) and (A.81), completes the proof.

A.13 Proof of Lemma 3.5

For

$$\phi(w) \tag{A.82}$$

$$= w\frac{\int_0^\infty (g+1)^{-p/2-a-1}\{1+w/(g+1)\}^{-(p/2+n/2+1)}\{g/(g+1)\}^b dg}{\int_0^\infty (g+1)^{-p/2-a}\{1+w/(g+1)\}^{-(p/2+n/2+1)}\{g/(g+1)\}^b dg},$$

note $w/(1+g) = -1 + \{1 + w/(1+g)\}$ in the numerator of (A.82). Then

$$\phi(w) = -1 + \frac{\int_0^\infty (g+1)^{-p/2-a}\{1+w/(g+1)\}^{-m+1}\{g/(g+1)\}^b dg}{\int_0^\infty (g+1)^{-p/2-a}\{1+w/(g+1)\}^{-m}\{g/(g+1)\}^b dg}, \tag{A.83}$$

for $m = p/2 + n/2 + 1$. Further

$$\frac{\{1+w/(g+1)\}^{-m+1}}{(1+w)^{-m+1}} = \left(\frac{1+w/(g+1)}{1+w}\right)^{-m+1} \tag{A.84}$$

$$= \left(1 - \frac{g}{g+1}\frac{w}{w+1}\right)^{-m+1} = \sum_{i=0}^\infty \frac{\Gamma(m-1+i)}{\Gamma(m-1)i!}\left(\frac{g}{g+1}\frac{w}{w+1}\right)^i.$$

Similarly

$$\frac{\{1+w/(g+1)\}^{-m}}{(1+w)^{-m}} = \sum_{i=0}^\infty \frac{\Gamma(m+i)}{\Gamma(m)i!}\left(\frac{g}{g+1}\frac{w}{w+1}\right)^i. \tag{A.85}$$

Then, by (A.83), (A.84) and (A.85), and using the hypergeometric function,

$$F(\alpha, \beta; \gamma; v) = \sum_{i=0}^\infty \frac{\Gamma(\alpha+i)}{\Gamma(\alpha)}\frac{\Gamma(\beta+i)}{\Gamma(\beta)}\frac{\Gamma(\gamma)}{\Gamma(\gamma+i)}\frac{v^i}{i!}, \tag{A.86}$$

we may express $\phi(w)$ as

$$\phi(w) = -1 + \frac{F(b+1, m-1; p/2+a+b; v)}{(1-v)F(b+1, m; p/2+a+b; v)},$$

where $v = w/(1+w)$. Further by Part 1 of Lemma A.20 below, we have

$$\phi(w) = -1 + \frac{m-1}{n/2+1-a+(p/2-1+a)G(v)}$$

$$= \frac{(p/2-1+a)\{1-G(v)\}}{n/2+1-a+(p/2-1+a)G(v)} \leq \frac{p/2-1+a}{n/2+1-a+b(p/2+n/2)},$$

where the inequality follows from Part 2 of Lemma A.20, where

$$G(v) = \frac{F(b, m-1; p/2+a+b; v)}{F(b+1, m-1; p/2+a+b; v)}.$$

This completes the proof of the first inequality of the lemma.

By (A.67), we have

$$\frac{w\phi'(w)}{\phi(w)} = 1 + mw\left\{ -\frac{\int_0^\infty (g+1)^{-p/2-2-a}\{1+w/(g+1)\}^{-m-1}\{g/(g+1)\}^b dg}{\int_0^\infty (g+1)^{-p/2-1-a}\{1+w/(g+1)\}^{-m}\{g/(g+1)\}^b dg} \right.$$

$$\left. + \frac{\int_0^\infty (g+1)^{-p/2-1-a}\{1+w/(g+1)\}^{-m-1}\{g/(g+1)\}^b dg}{\int_0^\infty (g+1)^{-p/2-a}\{1+w/(g+1)\}^{-m}\{g/(g+1)\}^b dg} \right\}.$$

By $w/(1+g) = -1 + \{1+w/(1+g)\}$, (A.84), (A.85) and (A.86), it follows that

$$\frac{w\phi'(w)}{\phi(w)} = 1 + m\left\{ \frac{\int_0^\infty (g+1)^{-p/2-1-a}\{1+w/(g+1)\}^{-m-1}\{g/(g+1)\}^b dg}{\int_0^\infty (g+1)^{-p/2-1-a}\{1+w/(g+1)\}^{-m}\{g/(g+1)\}^b dg} \right.$$

$$\left. - \frac{\int_0^\infty (g+1)^{-p/2-a}\{1+w/(g+1)\}^{-m-1}\{g/(g+1)\}^b dg}{\int_0^\infty (g+1)^{-p/2-a}\{1+w/(g+1)\}^{-m}\{g/(g+1)\}^b dg} \right\}$$

$$= 1 + m(1-v)\left\{ \frac{F(b+1, m+1; \ell+1; v)}{F(b+1, m; \ell+1; v)} - \frac{F(b+1, m+1; \ell; v)}{F(b+1, m; \ell; v)} \right\},$$

where $\ell = p/2+a+b$. By Part 1 of Lemma A.20,

$$\frac{w\phi'(w)}{\phi(w)} = (p/2+a)\frac{F(b, m; \ell+1; v)}{F(b+1, m; \ell+1; v)} - (p/2-1+a)\frac{F(b, m; \ell; v)}{F(b+1, m; \ell; v)}.$$

Further

$$(p/2 + a)\frac{F(b, m; \ell + 1; v)}{F(b + 1, m; \ell + 1; v)} - (p/2 - 1 + a)\frac{F(b, m; \ell; v)}{F(b + 1, m; \ell; v)}$$

$$= \frac{p/2 + a}{F(b + 1, m; \ell + 1; v)} + (p/2 + a)\frac{F(b, m; \ell + 1; v) - 1}{F(b + 1, m; \ell + 1; v)}$$

$$- \frac{p/2 - 1 + a}{F(b + 1, m; \ell; v)} - (p/2 - 1 + a)\frac{F(b, m; \ell; v) - 1}{F(b + 1, m; \ell; v)}$$

$$\geq (p/2 + a)\frac{F(b, m; \ell + 1; v) - 1}{F(b + 1, m; \ell + 1; v) - 1}$$

$$\geq \frac{(p/2 + a)b}{b + 1},$$

where the last inequality follows from Part 2 of Lemma A.20 below. This completes the proof of the Lemma's second inequality.

Part 1 of Lemma A.20 below is the the formula number 15.2.18 of Abramowitz and Stegun (1964). Part 2 of the lemma is essentially due to Maruyama and Strawderman (2009).

Lemma A.20   *1.* $(\gamma - \alpha - \beta)F(\alpha, \beta; \gamma; z) - (\gamma - \alpha)F(\alpha - 1, \beta; \gamma; z) + \beta(1 - z)$
$F(\alpha, \beta + 1; \gamma; z) = 0.$
2. For $-1 < b < 0$, $\beta > 0$ *and* $\gamma > 0$,

$$\frac{F(b, \beta; \gamma; v)}{F(b + 1, \beta; \gamma; v)} \geq \frac{F(b, \beta; \gamma; v) - 1}{F(b + 1, \beta; \gamma; v)} \geq \frac{F(b, \beta; \gamma; v) - 1}{F(b + 1, \beta; \gamma; v) - 1} \geq \frac{b}{b + 1}.$$

Proof   [Part 1] The ith component of the left hand-side is given by

$$\frac{z^i}{i!}\frac{\Gamma(\alpha + i)}{\Gamma(\alpha)}\frac{\Gamma(\beta + i)}{\Gamma(\beta)}\frac{\Gamma(\gamma)}{\Gamma(\gamma + i)}\left\{(\gamma - \alpha - \beta) - \frac{\alpha - 1}{\alpha + i - 1}(\gamma - \alpha)\right.$$
$$\left. + (\beta + i) - \frac{i(\gamma + i - 1)}{\alpha + i - 1}\right\}.$$

The term

$$(\gamma - \alpha - \beta) - \frac{\alpha - 1}{\alpha + i - 1}(\gamma - \alpha) + (\beta + i) - \frac{i(\gamma + i - 1)}{\alpha + i - 1}$$

is zero, which completes the proof of Part 1.
[Part 2] By $-1 < b < 0$,

$$F(b, \beta; \gamma; v) - 1 < 0 \quad \text{and} \quad F(b + 1, \beta; \gamma; v) - 1 > 0,$$

for all v. Then

$$\frac{F(b,\beta;\gamma;v)}{F(b+1,\beta;\gamma;v)} = \frac{1}{F(b+1,\beta;\gamma;v)} + \frac{F(b,\beta;\gamma;v)-1}{F(b+1,\beta;\gamma;v)}$$

$$\geq \frac{F(b,\beta;\gamma;v)-1}{F(b+1,\beta;\gamma;v)} \geq \frac{F(b,\beta;\gamma;v)-1}{F(b+1,\beta;\gamma;v)-1}.$$

Further

$$\frac{F(b,\beta;\gamma;v)-1}{F(b+1,\beta;\gamma;v)-1} = \frac{\displaystyle\sum_{i=1}^{\infty} \frac{\Gamma(b+i)}{\Gamma(b)}\frac{\Gamma(\beta+i)}{\Gamma(\beta)}\frac{\Gamma(\gamma)}{\Gamma(\gamma+i)}\frac{v^i}{i!}}{\displaystyle\sum_{i=1}^{\infty} \frac{\Gamma(b+1+i)}{\Gamma(b+1)}\frac{\Gamma(\beta+i)}{\Gamma(\beta)}\frac{\Gamma(\gamma)}{\Gamma(\gamma+i)}\frac{v^i}{i!}}$$

$$= \frac{b}{b+1}\frac{\displaystyle\sum_{i=1}^{\infty} \frac{\Gamma(b+i)}{\Gamma(b+1)}\frac{\Gamma(\beta+i)}{\Gamma(\beta)}\frac{\Gamma(\gamma)}{\Gamma(\gamma+i)}\frac{v^i}{i!}}{\displaystyle\sum_{i=1}^{\infty} \frac{\Gamma(b+1+i)}{\Gamma(b+2)}\frac{\Gamma(\beta+i)}{\Gamma(\beta)}\frac{\Gamma(\gamma)}{\Gamma(\gamma+i)}\frac{v^i}{i!}} \geq \frac{b}{b+1}.$$

This completes the proof of Part 2.   □

A.14 Proof of Theorem 3.15

Recall the estimator $\hat{\theta}_\alpha$ is $\{1 - \phi_\alpha(w)/w\}x$ where

$$\phi_\alpha(w) = w\frac{\int_0^\infty (g+1)^{-(\alpha+1)(p/2-1)-2}\{1+w/(g+1)\}^{-(\alpha+1)(p/2+n/2)-1}dg}{\int_0^\infty (g+1)^{-(\alpha+1)(p/2-1)-1}\{1+w/(g+1)\}^{-(\alpha+1)(p/2+n/2)-1}dg}.$$

As in (3.71) and (3.72), an integration by parts gives

$$(\alpha+1)(n+2)\int_0^\infty \frac{(g+1)^{-(\alpha+1)(p/2-1)-1}dg}{\{1+w/(g+1)\}^{(\alpha+1)(p/2+n/2)}} + \frac{2}{(1+w)^{(\alpha+1)(p/2+n/2)}}$$

$$= (\alpha+1)(n+2)\int_0^\infty \frac{(g+1)^{(\alpha+1)(n/2+1)-1}dg}{(1+g+w)^{(\alpha+1)(p/2+n/2)}} + \frac{2}{(1+w)^{(\alpha+1)(p/2+n/2)}}$$

$$= 2\int_0^\infty (g+1)^{(\alpha+1)(n/2+1)}\left\{\frac{(\alpha+1)(p+n)/2}{(1+g+w)^{(\alpha+1)(p/2+n/2)+1}}\right\}dg$$

$$= (\alpha+1)(p+n)\int_0^\infty \frac{(g+1)^{-(\alpha+1)(p/2-1)-1}dg}{\{1+w/(g+1)\}^{(\alpha+1)(p/2+n/2)+1}}$$

and

$$(\alpha+1)(p-2)\int_0^\infty \frac{(g+1)^{-(\alpha+1)(p/2-1)-1}dg}{\{1+w/(g+1)\}^{(\alpha+1)(p/2+n/2)}} - \frac{2}{(1+w)^{(\alpha+1)(p/2+n/2)}}$$

$$= 2\int_0^\infty (g+1)^{-(\alpha+1)(p/2-1)}\left\{\frac{w}{(g+1)^2}\frac{(\alpha+1)(p+n)/2}{\{1+w/(g+1)\}^{(\alpha+1)(p/2+n/2)+1}}\right\}dg$$

$$= (\alpha+1)(p+n)w\int_0^\infty \frac{(g+1)^{-(\alpha+1)(p/2-1)-2}dg}{\{1+w/(g+1)\}^{(\alpha+1)(p/2+n/2)+1}}.$$

Then

$$\phi_\alpha(w) = w\frac{\int_0^\infty (g+1)^{-(\alpha+1)P-2}\{1+w/(g+1)\}^{-(\alpha+1)(P+N)-1}dg}{\int_0^\infty (g+1)^{-(\alpha+1)P-1}\{1+w/(g+1)\}^{-(\alpha+1)(P+N)-1}dg}$$

$$= \frac{P-\mathcal{J}_\alpha(w)}{N+\mathcal{J}_\alpha(w)}$$

where $P = p/2 - 1$ and $N = n/2 + 1$ and

$$\frac{1}{\mathcal{J}_\alpha(w)} = (\alpha+1)\int_0^\infty \frac{dg}{(g+1)^{(\alpha+1)P+1}}\left(\frac{(1+w)(1+g)}{1+w+g}\right)^{(\alpha+1)(P+N)}$$

$$= (\alpha+1)\int_0^\infty \frac{dg}{(g+1)^{(\alpha+1)P+1}}\left(1-\frac{w}{1+w}\frac{g}{1+g}\right)^{-(\alpha+1)(P+N)}$$

$$= (\alpha+1)\int_0^\infty \frac{dg}{(g+1)^{(\alpha+1)P+1}}\sum_{i=0}^\infty \frac{\Gamma((\alpha+1)(P+N)+i)}{\Gamma((\alpha+1)(P+N))i!}\left(\frac{w}{1+w}\frac{g}{1+g}\right)^i$$

$$= (\alpha+1)\sum_{i=0}^\infty B((\alpha+1)P, i+1)\frac{\Gamma((\alpha+1)(P+N)+i)}{\Gamma((\alpha+1)(P+N))i!}\left(\frac{w}{1+w}\right)^i$$

$$= \sum_{i=0}^\infty \frac{(\alpha+1)\Gamma((\alpha+1)P)}{\Gamma((\alpha+1)P+i+1)}\frac{\Gamma((\alpha+1)(P+N)+i)}{\Gamma((\alpha+1)(P+N))}\left(\frac{w}{1+w}\right)^i.$$

For fixed i, the coefficient of $1/\mathcal{J}_\alpha(w)$,

$$\frac{(\alpha+1)\Gamma((\alpha+1)P)}{\Gamma((\alpha+1)P+i+1)}\frac{\Gamma((\alpha+1)(P+N)+i)}{\Gamma((\alpha+1)(P+N))}$$

$$= \frac{1}{P+i/(\alpha+1)}\frac{P+N+1/(\alpha+1)}{P+1/(\alpha+1)}\times\cdots\times\frac{P+N+(i-1)/(\alpha+1)}{P+(i-1)/(\alpha+1)},$$

is increasing in α. Then, by the monotone convergence theorem, we have

$$\lim_{\alpha\to\infty}\frac{1}{\mathcal{J}_\alpha(w)} = \frac{1}{P}\sum_{i=0}^\infty \left(\frac{P+N}{P}\frac{w}{1+w}\right)^i = \begin{cases}1/\left(1-\dfrac{(P+N)w}{P(1+w)}\right) & w < P/N,\\ \infty & w \geq P/N,\end{cases}$$

and hence

$$\lim_{\alpha \to \infty} \phi_\alpha(w) = \phi_{JS}^+(w) = \begin{cases} w & w < (p-2)/(n+2), \\ (p-2)/(n+2) & w \geq (p-2)/(n+2). \end{cases}$$

References

Abramowitz M, Stegun IA (1964) Handbook of mathematical functions with formulas, graphs, and mathematical tables, National Bureau of Standards Applied Mathematics Series, vol 55. Government Printing Office, Washington, D.C, For sale by the Superintendent of Documents, U.S

Geluk JL, de Haan L (1987) Regular variation, extensions and Tauberian theorems, CWI Tract, vol 40. Stichting Mathematisch Centrum Centrum voor Wiskunde en Informatica, Amsterdam

Maruyama Y, Strawderman WE (2009) An extended class of minimax generalized Bayes estimators of regression coefficients. J Multivariate Anal 100(10):2155–2166

Printed in the United States
by Baker & Taylor Publisher Services